elephas maximus

ALSO BY STEPHEN ALTER

Sacred Waters:
A Pilgrimage up the Ganges River
to the Source of Hindu Culture

Amritsar to Lahore:
A Journey Across the India-Pakistan Border

All the Way to Heaven:
An American Boyhood in the Himalayas

Renuka

The Godchild

Silk and Steel

Neglected Lives

STEPHEN ALTER

elephas maximus

A PORTRAIT OF THE INDIAN ELEPHANT

HARCOURT, INC.

Orlando   Austin   New York   San Diego   Toronto   London

www.HarcourtBooks.com

Picture Credits: Pages 7, 33, 63, 231, 257—Vivek Sinha. Pages 85,
111, 173—Stephen Alter. Page 145—anonymous artist from *The Indian Journal
of Art and Industry.* Page 199—anonymous artist from Martin Saller and
Karl Groning, *Elephants: A Cultural and Natural History.*

Quotation from *Arctic Dreams* by Barry Lopez. Charles Scribner's Sons,
© 1996, Charles Scribner's Sons. Reprinted by permission of The Gale Group.

Library of Congress Cataloging-in-Publication Data
Alter, Stephen.
Elephas maximus: a portrait of the Indian elephant/Stephen Alter.—1st ed.
p.   cm.
Includes bibliographical references.
ISBN 0-15-100646-6
1. Asiatic elephant.   I. Title.
QL737.P98A48   2004
599.67'6—dc22   2003017161

Text set in Columbus MT
Designed by Linda Lockowitz

Printed in the United States of America

First edition
A C E G I K J H F D B

For Rupak Roy
artist and sage
whose hermitage is visited by elephants

*Why no decorated elephant,
auspiciously adorned like a great mountain or black rain cloud,
promising a tranquil journey?*

—RAMAYANA

*Few things provoke like the presence of wild animals.
They pull at us like tidal currents with questions of volition,
of ethical involvement, of ancestry.*

—BARRY LOPEZ

contents

elephas maximus

prologue

a young bull elephant in Nagarhole National Park kicks at the dirt with his forefoot, then scoops up dust in his trunk and tastes it for salt. He is no more than fifteen years old, on the cusp of adolescence and moving apart from his mother's herd. Standing alone at the salt lick, he eyes us warily but does not back away. The surrounding forest of teak trees is a tawny green with broad leaves as big as the elephant's ears. It hasn't rained for several weeks and the winter grass and underbrush is dry, though a short distance away lies a forest pool surrounded by thickets of bamboo. The wild elephant is intent on finding salt, his head lowered in concentration and his trunk repeatedly probing the soil, then curling up to touch his pink tongue.

He stands over six feet tall at the shoulder but has yet to achieve the stature of a full-grown tusker. Part of him is still a calf, juvenile and clumsy, though his glands are beginning to squeeze the confusing juices of maturity into his veins. His head is rounded, his skull bulging under gray hide and a sparse bristle of black hairs. Short tusks protrude on either side of his trunk like ivory bayonets. All at once, something annoys him— the sound of our jeep, our whispering, the glint of sunlight off a camera lens. The elephant charges without warning, ears flared, tossing up a cloud of dust. He vents his anger in a frightening display, a brief confrontation between beast and man that takes place inside a dozen heartbeats. When our driver revs the jeep's engine and releases the clutch, the elephant quickly balks and swings around, his belligerence changing to fear. Turning tail, the young bull seems suddenly vulnerable, uncertain of his demeanor. We stop again as he retreats behind a stockade of trees. Facing us and now holding his ground, he listens to our nervous laughter, the excited murmur of human voices. Still anxious and aggressive, he watches us until we finally move away. With renewed bravado, the elephant charges after our departing jeep, emitting a triumphant squeal before reclaiming his position at the salt lick.

The fourteenth-century lexicographer Muhammad al-Damiri suggested that the elephant's tongue is upside down and if only it could be turned around this animal would be able to speak. Perhaps elephants might simultaneously begin to write as well, just as Ganesha is believed to have transcribed the *Mahabharata* epic using one of his tusks as a pen. Until such time, however, human beings are left to recount the life story of this species, even as we intrude upon the telling of the tale.

Elephants speak to us in many ways. Their gestures convey affection, curiosity, fear, and anger—human emotions for which an elephant has no true equivalent. Yet they are expressive animals who communicate with each other constantly in a herd, through noises that vary from subsonic rumbles and throaty gurglings to shrill trumpets of rage. The vocal range of an elephant is considerable and much of it lies below the pitch of human hearing. Listening to an elephant is like listening to the current of a river. Its breathing has a quiet rhythm that swells with waves of sound, a rippling exhalation culminating in a slow sigh, then a pooling silence before the next deep breath is drawn.

There is poetry in an elephant's eye, as it watches us with calm intensity. Compared to the massive bulk of its head and body, an elephant's eyes are tiny, set within a whorl of wrinkled skin, its pupils dark and moist. The impulse to anthropomorphize these beasts is irresistible and even scientists succumb to the temptation, giving the subjects of their fieldwork human names and describing their behavior as one might describe our own society. In ancient Sanskrit texts, the character and psychology of elephants is compared to human traits, the ideal animal being one "who is gentle in all his feelings, and free from vice . . . that elephant the noble sages call one of perfect sensitivity." Yet even as we attempt to define the elephant in our own terms, we must recognize that there is a great deal about its behavior that we will never understand.

One morning in Delhi I watched two tame elephants standing side by side—a tusker and a cow. They leaned against each other, rubbing flanks and exchanging caresses with their trunks. The bull was at least a foot taller, with blunt tusks trimmed off at the ends, to avoid his causing injury. Both animals were covered in dust and their legs were caked with dry mud, but their faces and ears bore faint traces of floral designs and smudged curlicues. Decorated by their owners and hired out for the night, these elephants provide an auspicious presence at wedding receptions. The celebrations were long over, though, and in the bright morning sunshine the two elephants looked like a pair of exhausted partygoers supporting each other and enjoying a residual moment of romance. The female nuzzled the male, who flapped his ears contentedly as she pressed her head against his shoulder. They swayed together for a few moments, caught up in their own internal tempo, ignoring everyone around them. Then with sensuous intimacy, the tusker curled his trunk and brushed the female's cheek before he put the end in her mouth. For several seconds she held the tip of his trunk between her lips, as close to a kiss as an elephant can come.

In this book, I have tried to tell the elephant's story in India through myth, art, and literature, as well as something of its biology and natural history. It should be clear that I am not a scientist like those who spend years studying the elephant and its habitat. Nevertheless, I have been privileged with a lifelong fascination for the forests of India and my encounters with elephants go back into childhood. What I have written is essentially a homage to *Elephas maximus* and an alarm call for its protection. More than any other animal, elephants represent the indomitable force of nature as well as its vulnerability. To watch an elephant break branches from a tree is to know how something that has taken years to grow can be ripped apart in minutes, but this is nothing compared to the irreversible destruction that man has caused.

The sequence of chapters is based on a series of journeys I made to different parts of India in 2001–2002. It is roughly chronological, recounting my experiences at national parks and various elephant venues. Though there are a number of other places where elephants are found in India, this itinerary allowed me to draw together stories and observations from each of the major regions of the subcontinent. I began in midwinter, with a visit to Corbett National Park in northern India, then traveled over 2,000 kilometers south to Mysore and Nagarhole National Park. From there I continued into the southernmost states of Tamil Nadu and Kerala, where I visited Theppakkadu Elephant Camp in Mudumalai Sanctuary and Periyar Tiger Reserve, as well as the elephant stables at Guruvayur Temple. In early summer, I returned to north India and spent time at Rajaji National Park, along the upper reaches of the Ganga, or Ganges, River. Between these visits I traveled through Delhi several times and during early September, at the end of the monsoon, I attended the Ganesha Chathurthi festival in Mumbai on India's western coast. Side trips took me to Ajanta and Ellora, in central India, to search for images of elephants in ancient art, as well as to the fortified city of Kotah in Rajasthan, where elephants once played a unique role in royal festivals. In November, I attended the annual Sonepur Mela, a cattle fair in northern Bihar where elephants are bought and sold. The book concludes with a journey to the northeastern state of Assam and Kaziranga National Park.

A few words of explanation may be necessary regarding terminology. *Elephas maximus* is now commonly referred to as the "Asian elephant," a name that better describes its geographical distribution than the "Indian elephant." However, I have used the latter in this book because my focus is elephants in India. Wherever possible I have tried to include the current names of cities and places, i.e., Mumbai instead of Bombay, Mamallapuram instead of Mahabalipuram. Within a historical context, however, it has sometimes been necessary to revert to the original name. Some readers will complain about inconsistencies in my transliteration of words or names from Indian languages. I have tried to balance common usage with scholarly spellings, which are often more confusing than helpful. One important omission, which may disappoint some readers, is that I have avoided descriptions of circus elephants. The reason is simple: Training elephants to do circus stunts is a cruel and painful business. I see no reason why animals need to be subjected to this kind of treatment, even if it makes children smile. Elephants provide more than enough amusement and spectacle without having to ride tricycles, stand on their heads, or wear silly costumes.

Observing elephants in the wild is the only way to begin to understand their complex nature and Nagarhole National Park is one of the best places in India to view these animals in their natural habitat. Along the banks of the Kabini River, which runs through the sanctuary, I watched one of many different herds browsing on winter grasses. The extended family of adult females and calves stood in sunlight and shadow, fifteen to twenty animals altogether. Their solid gray shapes moved slowly along a leafy perimeter of bamboo. Dozens of trunks swung back and forth, as they plucked the grass in unison, then dusted it against their knees before each animal took a mouthful. Though the elephants were directly in front of me, less than fifty meters away, there was an abstract quality to the scene—as if I were looking at them through a watery lens. The graceful movement of their huge bodies and the contrasting light and shade of a winter afternoon made it seem like a painting in which the colors were still wet, a fluid landscape with creatures blending into one another, almost as if it were a single animal.

The society of elephants, their interaction in a herd, the way in which they move together suggests kinship and solidarity. They touch each other continually, brushing flanks and wrapping trunks, coddling the young and seeking physical reassurance from their elders. Even as every gesture and instinct of the species reflect the biological imperatives of survival, there is something else that seems to occur within a wild herd of elephants, a much more subtle and intangible purpose—an almost spiritual communion. To the human eye elephants may seem immeasurable, yet we share the same dimension, a sense of common ground.

I

winter sanctuary

January 3, 2002

We had spent eight hours that day in pursuit of elephants, driving from early morning to late afternoon along the dusty roads that circle through Corbett National Park. Accompanying me was a close friend, Ajay Mark, as well as our driver, Sanjay Singh, and a guide assigned by park authorities, M. S. Negi. Our search had taken us across open grasslands, up boulder-strewn riverbeds, and through heavy jungle. Despite their size, elephants have a way of disappearing in the forest and none of these huge beasts had revealed themselves. We saw plenty of other animals, including four kinds of deer—spotted cheetal, barking deer that have rusty red coats, sambar with shaggy brown manes, and hog deer that are a dull gray color and duck into the underbrush at the first growl of a car's engine. Sounders of wild boar rooted about near the roadside and langur monkeys with silver fur sprang from branch to branch overhead. But there were no elephants that we could see.

Signs of them were everywhere—broken tree limbs and mounds of dung like fibrous clumps of drying mulch. Obviously, a herd of elephants had been nearby in recent days, browsing over the grasslands and along the leaf-tangled margins of the jungle. Near the river we could see the circular prints of their feet in the mud and we found a place where one of them had wallowed near the shore, leaving a broad depression partly filled with water. Marks from creases in the elephant's skin were etched in the drying clay and looked like giant fingerprints.

I was told that only two days earlier a family of seven elephants had wandered up to the gate of the park headquarters at Dhikala. "They were right here," said Negi, pointing to the high grass encircling the camp. "All night we could hear them feeding."

Now the elephants were gone, retreating into the forest like the shadows of trees. Seated on the veranda of the rest house, I could see the foothills across the river fading into darkness. Stars were coming out and Venus shone overhead like a beacon, so bright it seemed artificial. The electricity at Dhikala had gone off and only a few kerosene lanterns glowed in doorways of the camp. My frustration at not having seen a wild elephant was tempered by a comforting sense of separation from the world of jet airliners, digital communication, and the so-called war on terror that dominated recent headlines. Though Dhikala is a large camp, housing over a hundred people, I had a feeling—particularly with the lights off—that this was a world in which human beings were thankfully outnumbered. The venality and violence of civilization, our petty jingoism and material conceits, counted for little under a black sky, pierced by the innumerable sparks of distant galaxies.

Alone with my thoughts, I imagined those shadowy elephants moving about under the cover of night. Their invisible presence seemed as constant and fluid as the nearby river, which had also disappeared. The elephants were elusive, but I knew that somewhere close at hand, certainly nearer than the stars, a herd was roaming through the forest. In my mind they grew larger and larger, taller than the trees, taller than the ridges on either side of the valley. Surrounded by darkness, I had no sense of scale, no skyline against which to measure my imagination. For me the elephants could be as huge as mountains or as small as specks of dust in my eye. Yet all I wanted was to see them, to witness the movement of their bodies and trace the swaying of their trunks. The footprints, the shattered branches and piles of dung weren't proof enough. I needed to reaffirm with my own eyes that they were here.

To fully appreciate the elephant it is essential to understand the habitat in which it lives. Few places in India are as beautiful as Corbett Park, where the Ramganga River emerges from the lower Himalayas and cuts a course that runs parallel to the foothills for thirty kilometers before spilling out into the plains. The Patlidun, literally "narrow valley," lies at the heart of this sanctuary. Thirty years ago a dam was built downstream, creating a reservoir, but above Dhikala the river remains pristine and undisturbed, flowing between densely wooded hills and bordered by an undulating mosaic of rounded rocks. The Ramganga is the vital artery of the park, a river that teems with fish like the mahseer and the goonch, sometimes referred to as a "freshwater shark." Two kinds of crocodiles are also found in this river. The mugger is more common, with a broad, flat snout, while gharial crocodiles have long, thin jaws and the males grow a bulbous nose.

Immediately upon entering the park one has a feeling of enclosure as the road passes through thick sal forests, the broadleafed trees forming a cavern of branches overhead. In places there are clearings, revealing a muddy water hole or a brief glimpse of rugged foothills. The road is scored by dozens of dry streambeds that flood during the monsoon, though in winter these are littered with boulders and fringed with tall grass. The park is one of the last sanctuaries for the elephant and a place that preserves India's forest heritage, which has gradually been eroded by the felling of trees, the spread of agriculture, and the encroachment of human population.

With a park area of 520 square kilometers and contiguous reserve forests stretching over 2,000 square kilometers, Corbett

National Park provides a refuge for many different species of wildlife, particularly the tiger. Roughly 350 elephants live here and until recently these were safe from poaching. However, in the winter of 2000–2001 eight bulls were killed in the park and their tusks hacked off. There are differing accounts of the method by which these animals were slaughtered. Newspaper articles reported that jagged pieces of metal and sharp nails were hidden inside balls of dough that were left in the path of the wild tuskers, who swallowed them whole. Once the sharp objects were ingested they lacerated the stomach and intestines of the elephants, who slowly bled to death. Later investigations suggest that the elephants were killed by poison arrows, implying that the poachers came from the northeastern states of Assam and Arunachal Pradesh, where this method of hunting is common.

Reading newspaper articles about the elephant deaths, I couldn't help but feel a personal sense of violation and anger at the way in which the sanctuary had been desecrated. When I was a child, our family made annual trips to the park during winter holidays, and we would spend a week or more at one of the forest bungalows along the banks of the Ramganga. These trips provided some of my formative memories of the Indian jungle, driving in an open jeep at dawn, spotting animals in the mist. Corbett Park is where I saw my first tiger in the wild, as well as leopards and crocodiles, those shy predators that excited my imagination. It was here, too, I saw wild elephants for the first time, a memory that remains vivid after thirty-five years.

My father had stopped the jeep at the side of a forest track, where we could see that elephants had recently crossed. Over the sound of the idling engine I heard them snapping branches as they fed, less than a hundred feet away. The foliage shook and trembled but the elephants were hidden behind a curtain of leaves. We waited for ten minutes, then cautiously got down from the vehicle and began to enter the forest. Immediately one of the elephants stepped into view, a large female holding a switch of leaves in her trunk. A pair of tiny, somber eyes looked down at us with matronly disapproval. The gray bulk of her body swayed from foot to foot with the imposing indignation of a stern head-mistress. As she waved her switch accusingly in our direction, I felt vulnerable and strangely guilty for sneaking about the jungle. Each of us held our breath and got ready to dive back into the jeep, but just when I expected the elephant to charge, she stuffed the leaves into her mouth and turned away complacently, as if she had convinced herself that we no longer merited her displeasure.

Years later, reading about elephant poaching in Corbett Park, I felt an immediate impulse to go back. More than any-thing I wanted to locate those animals again and confirm their presence. The cruel deaths of those tuskers seemed to threaten the sanctity of the Ramganga valley. Beyond the outrage and re-vulsion that I felt was an enigmatic sense of loss and separation, the knowledge of extinction.

Within a few months Corbett Park was in the headlines again. This time a forest ranger had been shot dead and three guards wounded when they tried to apprehend a gang of poach-ers. As before there was a brutal and mindless savagery to these deaths that seemed so alien to my memories of the protected forests and peaceable grasslands. This time the victims were not elephants but men, equally innocent, their lives cut short by human predators acting out of greed and cowardice. It seemed as if even this tiny strip of jungle, a final remnant of the sprawling forests that once bordered the Himalayas, was not safe from the violence of this new millennium.

Flagship Species

At the main gate of Corbett Park stands a recently erected me-morial, a cairn of polished river rocks framing a black marble plaque in remembrance of forest department officials who have lost their lives protecting wildlife. There is also a bust of Jim

Corbett, for whom the park is named. He was a British hunter and naturalist, who was born in India and spent much of his life in the nearby village of Kaladunghi. Famous for shooting man-eating tigers and leopards in the 1920s and 1930s, Corbett has become a central figure of India's jungle lore. He published six books about his adventures in the Himalayan foothills. Unlike most tales of shikar written during the British Raj, Corbett's hunting stories are remarkably free of the bravado and bluster of empire, conveying a deep understanding of natural history and a personal affinity for the people of Kumaon. In 1936 he helped establish the wildlife sanctuary that was initially named after Malcolm Hailey, then governor of the United Provinces. It is significant that this park was renamed in Corbett's honor after independence, when many colonial names were being erased. In 1973 Project Tiger was launched—a conservation initiative that focuses on a single endangered species—and Corbett Park was designated as a primary tiger reserve.

The conservation of wildlife in India is inevitably linked to the history of shikar. Virtually all of the sanctuaries and national parks were once shooting blocks, where either the forest department or the rulers of princely states protected animals for the sake of blood sport. After a total ban on hunting was imposed in 1972, the role of state forest departments changed significantly, though the methods of wildlife management retain some of the traditions of shikar, particularly the pursuit of wildlife on elephant back. The two animals most closely associated with big-game hunting in India were the tiger and the elephant. Left to themselves, both species generally avoid each other in the wild. One being a carnivore, the other an herbivore, tigers and elephants seldom come into conflict, except in rare instances when tigers prey on elephant calves. However, human beings have linked these two species through the rituals of hunting, generating myths of enmity, and celebrating the valor of a hunter shooting a tiger from the back of an elephant.

From the beginning of the sixteenth century onward, Mughal emperors used elephants in shikar, which served as a favorite subject for artists who illustrated their memoirs. It is even possible to follow the gradual improvements in weaponry through a chronology of these paintings, as bows and spears give way to muskets. Ideally suited to negotiating the rugged terrain of Indian forests, elephants safely carried the hunters through grassland and jungle, providing a secure platform from which a shikari took his aim. In the course of a hunt, elephants were occasionally forced to fend off attacks by wounded animals, and in miniature paintings tigers or lions are sometimes shown leaping onto a tusker's head in a desperate attempt to kill the hunters. Though situations like this were probably rare, the ferocity of the encounter was embellished by the artist's brush.

After the Mughal Empire disintegrated, many smaller kingdoms carried on these traditions, maintaining stables of hunting elephants and forest preserves. Rulers of the princely states of Rajasthan were particularly enamored of shikar, which also features prominently in miniature paintings from their courts. More than anyone, however, the British took pleasure in shikar, both as a pastime and as a performance. The white hunter became symbolic of imperial adventure and the heroics of colonial shikaris, who "dispatched," "bagged," "destroyed," or "disposed of savage predators." Much of this hunting continued to take place from the backs of elephants, and whenever dignitaries came to India—whether they were members of the British royal family or other VIPs—a tiger shoot on elephant back was considered de rigueur. Various maharajahs, who had stables of trained elephants and easily accessible hunting preserves, hosted these elaborate shikars. Sometimes as many as 300 elephants would be lined up to beat a jungle, moving in unison through the high grass and flushing whatever game appeared in front of them. As soon as a tiger was located—usually trying to slink unobtrusively out of sight—the elephants converged in a ring until the distinguished visitor was able to fire a shot. In many ways, this kind of shikar was more theatre than sport.

Queen Victoria never visited India or participated in a shikar, but each of her successors engaged in this ritual—Edward VII (as Prince of Wales) in 1876, George V in 1911, following his Coronation Durbar, and Elizabeth II in 1961. The Queen and her consort, Prince Philip—later president of the World Wildlife Fund—were hosted by King Mahendra of Nepal in a royal shoot in the Terai jungles of Chitwan, east of Corbett Park.

For those who were not born into the upper ranks of the aristocracy, shikar served as a means of improving social standing within the snakes and ladders of Anglo-Indian society. The colonial class structure was almost as rigid as the Hindu caste system, but hunting prowess allowed for a certain amount of upward mobility, through the assertion of masculine ideals. Jim Corbett serves as one of the best examples of the way in which shikar elevated an individual's status. As the country-born son of a postmaster in Nainital, Corbett hardly had an acceptable pedigree. It was only after his success in killing man-eaters that he was admitted to the upper ranks of the British Raj.

Many colonial hunters approached shikar as a competition, in which they sought to outdo each other by shooting as many animals as they could. In addition to tigers and other predators, elephants became prized trophies and some hunters declared them to be "the most dangerous game." Before the arrival of the British, elephant hunting in India usually involved capturing these animals rather than killing them. Even among shikaris of the Raj there were many who felt that "elephants were something one shot from, not shot at." Nevertheless, solitary tuskers, often misrepresented as rogues, were judged fair game, and shikaris boasted about "bringing down" a charging elephant with a well-placed bullet to the brain.

Alongside the indiscriminate slaughter of wildlife there was a growing awareness of nature's limits. Though hunting continued in India throughout the Raj, often on an excessive and ruthless scale, many colonial officials sought to control and regulate shikar. In his introduction to *The Great Indian Elephant Book,* D. K. Lahiri-Choudhury describes how attitudes toward hunting changed, particularly concerning elephants. Reserve forests were established in various parts of India from 1860 onward, and by 1871, elephants were declared a protected species in the Madras Presidency. In 1879 the All-India Elephant Preservation Act was put into force, but as Lahiri-Choudhury points out, "More than 'preservation,' the 1879 Act was meant to establish government monopoly of this important and strategically vital natural resource." Captured elephants were used by the government as working animals and also sold to raise revenue.

Though initial efforts at wildlife conservation under the Raj were essentially motivated by a desire for commercial gain, attitudes toward shikar began to change. Hunting became somewhat more selective, and success was not always measured by the number of animals shot but by the skill and perseverance of the shikari, as well as his understanding of the forest. The literature of shikar, which often reads like a catalogue of carnage, began to change as well. Writers like Corbett and F. W. Champion, who preferred the camera to the rifle, focused on India's natural heritage and warned about the consequences of unregulated hunting and the destruction of habitat. In *Man-eaters of Kumaon,* Corbett writes: "A tiger is a large hearted gentleman with boundless courage and when he is exterminated—as exterminated he will be unless public opinion rallies to his support—India will be the poorer by having lost the finest of her fauna." Like Champion, Corbett turned to wildlife photography toward the end of his life and recorded some of the first cinematic images of tigers in the wild.

In many ways, Jim Corbett represents the antithesis of tiger hunting from the back of an elephant. The man-eaters he killed were pursued on foot and he usually hunted alone. Once his books became bestsellers, though, Corbett was often enlisted to accompany dignitaries into the jungle. "The Talla Des Man-eater," one of his last stories, begins uncharacteristically with a hunt on elephant back at a place called Bindukhera, not far from Corbett Park.

On a February morning in 1929, Corbett and his party mounted a contingent of seventeen elephants. Over the course of that day the hunters killed quail, partridges, jungle fowl, peacocks, florican, hog deer, and a leopard. But the incident that Corbett focuses on took place around noon, when a ground owl was flushed by one of the elephants. The hunters held their fire but the owl was immediately attacked by a peregrine falcon. All eyes were fixed on the two birds, as they engaged in a desperate aerial chase. Corbett signaled for the line of elephants to stop while the owl circled up into the sky and finally eluded the falcon by disappearing into a cloud. At this point the shooting party erupted in cheers. The irony did not escape Corbett's attention, as he writes: "The reactions of human beings to any particular event are unpredictable. Fifty-four animals had been shot that morning—and many more missed—without a qualm or the batting of an eyelid. And now, guns, spectators, and mahouts were unreservedly rejoicing that a ground owl had escaped the talons of a peregrine falcon."

Wherever the line may be drawn between exploitation and conservation, the persistent and paradoxical memory of shikar is likely to remain a part of wildlife management in India for some time. Though the tiger and the elephant have been released from their roles as combatants in shikar, they remain the "flagship species" of conservation. In many sanctuaries like Corbett Park, forest department elephants still carry visitors into the jungle to search for tigers and other wildlife. Fortunately, the objective now is sighting animals rather than shooting them.

Melancholy

In the first gray light of a winter dawn we climb onto the back of an elephant and descend into the grasslands bordering the Ramganga. The line between sky and land is barely distinguishable, blurred shades of blue smudging the profiles of the mountains. Layers of mist lie over the river like strips of crumpled muslin. Though it hasn't rained for weeks, the air is cold and damp. Our elephant, a young female named Mohini, moves silently as she makes her way down to the riverbed. A pair of jackals loiter at the edge of the embankment but she ignores them and wades into a shallow pool of water. There is still no sound, not even the slightest splash, as her huge feet sink into the soft mud. Only when Mohini pauses to drink do I hear a faint gurgle as she sucks the water into her trunk, then squirts it between her lips.

Shrouds of moisture obscure the grasslands where the Ramganga splits into three separate streams. The valley has been drained of color and all of its shapes are undefined. The grass is white with frost, as if coated by a rime of salt. Seated atop the elephant, we are level with the mist, levitating upon a cloud. Looking up, I can just make out the dark contours of the foothills and the high bank at Dhikala, which is still in shadow, though all around us the world has been erased. My breath condenses each time I exhale, as if the mist were passing through my lungs.

By the time we reach the main branch of the river, my hands are numb with cold. Above the water I can feel the temperature drop even further, though Mohini doesn't seem to notice as she wades across. Slowly the sky begins to brighten but when the sunrise comes, it is sudden. A saffron glow at the head of the valley congeals into an amber ball. The mist turns golden and within seconds it begins to lift, like cobwebs disintegrating in a breeze.

There is some movement now, a white-necked stork flying just above the water, its wings slicing through remnants of the mist like feathered shears. A cormorant bobs to the surface of the Ramganga for a few seconds, then vanishes back into the glassy current. Cheetal are grazing in the distance and raise their heads as we approach. The first tentative bird calls can be heard, like the instruments of an orchestra being tuned—the trilling of warblers, a crow pheasant's plaintive cry, and the flustered shrieks of jungle babblers. The river is now audible, too, its current ruffled by submerged shelves of rock. An odorless dawn gives way to subtle smells, the sour scent of mud along the riverbank, the fragrance of crushed grass, the perfume of chewing tobacco that Muhammad Nayyab, Mohini's mahout, tucks into his mouth.

More colors have appeared in the landscape and the river turns a gilded green. As the frost melts, the dry grass reveals warm hues of ocher, red, and brown. I begin to notice flowers, blue tufts of ageratum and orange lantana. A flash of turquoise catches my eye as a kingfisher takes flight, while the variegated green shapes in the forest become clearer, trees forming out of shadows, mountains unfolding into leafy ridges.

All this time I have been scanning the valley for wild elephants, hoping to see their shapes emerge from the mist. As the edges of the forest are flooded with sunlight, I strain my eyes, expecting the shadows to reveal a gray body with flapping ears and waving trunk. Each movement signals the possibility of elephants but resolves itself into something else—a lone boar turning to face us with suspicion, a sambar doe flicking her ears in alarm, a kestrel rising out of the grass and hovering overhead like a fluttering chevron.

Distances and time are difficult to measure when you are riding on the back of an elephant. At first it seems to take forever to travel from one place to the next; then all at once you are there, as if transported in a dream. Mohini moves steadily through the tall grass, flushing partridges and red jungle fowl. At the far side of the valley we come to a section of grass that has been burned by the forest department. Over the next two months these fires will continue to be lit, promoting fresh growth and avoiding more serious fires in summer. Mohini steps cautiously around a smoldering tussock and takes a shortcut through a swathe of black ash. Muhammad Nayyab whispers to me over his shoulder: "The best season to see elephants is summer when most of the water holes in the forest dry up. At that time of year there will be herds everywhere, feeding on the new grass and bathing in the river. Come in May or June and you'll have no trouble finding elephants."

Winter may not be the ideal time of year to observe animals in Corbett Park, but the climate is much more pleasant than in summer, with cooler temperatures and fewer mosquitoes. Muhammad Nayyab nudges Mohini forward with his toes pressed behind her ears. She pauses to snap off a trunk full of leaves, then enters a patch of scrub jungle. Two vultures and a pariah kite sit on the branches of a nearby semal tree, signaling the possibility of a tiger's kill. If we were on foot, it would be impossible to see more than a few inches ahead of us but from the back of the elephant we have a grandstand view. After a hundred meters we come upon the tiger's kill, hidden within a tangle of grass. It is a young cheetal stag, his stubby horns in velvet. The lower half of the carcass has been eaten, from the rib cage down, but the upper portion is intact, one foreleg raised, as if the stag were still attempting to escape.

The tiger has fed on the cheetal during the night and then moved away from his kill. No doubt he has hidden himself nearby to sleep off his meal. Though Mohini makes three circuits of the area, we cannot locate the tiger. Surprisingly, there are several cheetal grazing within twenty feet of the kill. The deer are wary, though, and with our approach they take flight, porpoising through the grass in fluid leaps and bounds.

When I was a boy, one of the high points of our visits to Corbett Park was the elephant rides at Dhikala. My brothers and I would scramble up the steps to a cement platform built for mounting the elephants. As they approached from the stables at the edge of the camp, the three of us waited eagerly. Our favorite elephant was named Malan Kali, whom we had rechristened Melancholy because of the sad yet patient look in her eyes. She was the elephant we always asked to ride, perhaps on account of her name or because she was the largest.

Our whole family could sit on Melancholy's back, though my mother preferred to stay on solid ground because the rocking motion of an elephant's stride made her seasick. Huddled together in the cold, my brothers and I would clutch at the legs of the upturned cot that served as a howdah. Setting out into the Patlidun, I felt as if we were riding aboard a raft, adrift on a rustling sea of grass. We watched peacocks scuttling down the trail ahead of us and listened for the staccato cries of black partridges, which sound like Morse code. Each of us would point, or nudge one another if we spotted a deer or boar.

More exciting than the animals we saw, however, was the feeling of freedom in the forest and the elephant's ability to go anywhere, pushing through obstacles that seemed impassable. Without pausing, Melancholy trampled through barricades of lantana or brushed aside an overhanging branch. The elephant was able to climb steep hills and embankments, as we lurched from side to side, almost toppling off her back. Descending to the Ramganga, Melancholy waded across the river, its current parting at her knees. She was surefooted on the slippery stones and never seemed to stumble even when she picked her way through a maze of fallen tree trunks.

Though an elephant may appear flat-footed, it actually walks on its toes. The soles of its feet are splayed out like callused saucers, but the bones inside extend downward at an angle and the heel is thickly cushioned with a padding of fleshy tissue that absorbs the weight of three or four tons. This is one of the reasons an elephant can walk as swiftly and silently as it does. Despite its agility, however, the one thing an elephant cannot do is jump, though its stride is long enough to cross most obstructions.

As Melancholy moved through the forest, her trunk directed her, testing the air, siphoning water from a stream, or sampling leaves and grass along our route. My brothers and I would lean forward to watch her stuff these into her mouth, intrigued by the pink wedge of her tongue and her jaws that moved from side to side as she chewed. We also delighted in the way she farted, pinching our noses and snickering into our fists whenever we heard the sputtering sound, like air escaping from the bladder of a football. Whoever was seated at the back could see the elephant ceremoniously raise her tail and expel huge clods of dung the size of cannonballs.

At the end of our rides we would reluctantly crawl off Melancholy's back and run to fetch the stalks of sugarcane that we had brought for her as treats. She took them in her trunk, then fed herself with a deliberate slowness, as if savoring each bite, the juice dribbling from her mouth. Gathering my courage I reached out to touch her trunk and felt the reassuring roughness of her wrinkled hide, textured like the bark of a tree.

Elephants can live beyond seventy, but I did not expect to find Melancholy when I went back to Corbett Park this time. The mahouts at Dhikala told me that she had died ten years ago. They seemed pleased that I recalled her name, though they insisted it was Malan Kali. For me, however, there will always be a hint of sadness that I remember in her eyes—not because of captivity or abuse, but because of something else, a sense of loss and separation. Perhaps it wasn't her melancholy at all but a reflection of my own, a wistful loneliness that surrounded me as we drifted away upon a sea of grass.

Tamed by the Wind

Anyone who has stood next to an elephant can testify to its enormous size, as well as an immediate sense of affinity that human beings feel for their fellow mammal. In the company of elephants we recognize our own place in nature and respond with awe and affection. Despite the obvious differences in physical form, there are very few animals with which we share such empathy. Looking at an elephant's hide up close is like seeing our own skin under a microscope, the patterned texture of ridges and furrows, the hairs and wrinkles magnified. Though these huge creatures have been captured and trained by man for centuries, they still carry themselves with an aloof dignity and retain something of their wild nature.

Most tame elephants in India have two or three handlers. The mahout is responsible for the elephant's training, its day-to-day maintenance, feeding, and health. Assisting him are the charrawallas (fodder cutters), whose job it is to collect fresh leaves and grass from the forest, keep the stables clean, and give the elephant its daily bath. Charrawallas usually work as apprentices, aspiring to become mahouts themselves once they have gained experience.

On our second day at Corbett Park, Ajay and I watched three elephants being given a bath in the river. This is a time-consuming process, for the elephant has to lie down in the water as the charrawalla scrubs its hide to clean the animal and protect it from fungal infections as well as insects and other parasites. Though an elephant's hide can be over an inch thick in places, it is extremely sensitive and injuries or lesions easily fester unless properly treated. In captivity, many skin problems arise because of abrasions from the ropes and chains that are used to restrain an elephant, as well as the loads it carries on its back. A stable not properly cleaned and equipped with adequate drainage can cause infections and foot rot. In the wild, an elephant is less susceptible to these ailments but regularly bathes itself and throws mud and dust on its skin as protection.

Though the water was cold, the elephants seemed to enjoy being bathed, and they lolled on their sides in a shallow section of the river, as the men rubbed and massaged them from tail to trunk. When given commands to roll over, the animals complied and obediently flapped open their ears so that these could be scrubbed both back and front. This was followed by a pedicure, as the charrawallas scoured each toenail and scraped the callused soles of the elephants' feet.

Rolling sideways, with a reluctance that suggested modesty, the elephants parted their hind legs, revealing the pink labial folds of their genitals. All three of the elephants were females and between their forelegs they had small breasts that almost looked human. The men splashed water on these and rubbed the elephants' teats with their hands. Though the charrawallas worked efficiently, with practiced ease, there was an intimacy about the bath so that taking pictures with my camera I began to feel like a voyeur. For the elephants there seemed to be a kind of erotic pleasure to these rituals, as they lay in the shallow water and waved their trunks about in sensuous delight. They allowed the men to walk barefoot all over their bodies, as their skin was scrubbed with pumice stones. It was obviously hard work for the charrawallas, but there was a gentleness to their touch, the kind of physical empathy that men and animals often have when they live in close proximity—a purely tactile understanding. Later, after returning to the stables, these same men would anoint the elephants' foreheads with oil and draw floral patterns around their eyes with colored chalk, as if applying creams and makeup.

When the elephants finally rose from the river, their skin was gleaming black and the water spilled off them like streams running down the sides of a mountain. Mohini, whom we had ridden on that morning, seemed to glow, the pale pigmentation of her trunk and ears contrasting with her dark hide.

The largest of the three elephants was called Pawanpali, which means "tamed by the wind." A lanky, long-legged cow, she stood over seven-and-a-half feet tall. According to her mahout, Nazir, Pawanpali was about forty years old. A thin, wiry man with hair dyed jet black, Nazir is one of the senior mahouts at Dhikala. He has worked at Corbett Park for thirty years. Like most elephant handlers in northern India, Nazir is a Muslim.

Mahouts and charrawallas live in a line of quarters beside the elephant stables. Pawanpali and Mohini are owned by the forest department and the mahouts are employees of the park, but Nazir and the other handlers take a proprietary interest in their elephants, especially when it comes to health and medical care. They are suspicious of government veterinarians and prefer to use their own herbal remedies.

Nazir told me that five years ago Pawanpali gave birth to a calf and nearly died of complications afterward. It was her third pregnancy, though the first two resulted in stillbirths. The period of gestation for an elephant is twenty to twenty-two months. This time Pawanpali carried the calf to full term and had a successful delivery, but a few days later her abdomen became distended.

"She swelled up so much that her stomach touched the ground," said Nazir. "The park director sent for a veterinarian and when he came to examine her he said there was no hope and she was going to die. They gave her some injections and other medicines but these didn't help. The doctor knew nothing about elephants and ordered buffalo milk for the calf, which I knew it couldn't drink. Against his orders I made the baby nurse on Pawanpali's milk, even though she was very sick.

"Then I sent word to an old mahout, who used to be a friend of my father. He came here immediately and brought with him our own kind of medicine, made from herbs and roots. Together the two of us cut open the abscess in her stomach and all of the pus came flowing out. There was so much of it that the ground was flooded. After that we cleaned the wound and put a poultice on it. Even then we forced the baby to drink its mother's milk for this made both of them stronger. The forest officers and the government veterinarian could not believe that Pawanpali survived. Afterward they commended me for saving her life but in the beginning they wouldn't listen. You see, a mahout knows these things about an elephant. It is my life as much as hers."

Elephants rarely breed in captivity, but at a forest camp like Dhikala, where tame females often encounter wild bulls while grazing in the forest, there is a much greater possibility that they will find a mate. Unlike most domesticated animals, elephants retain many of their instincts from the wild and can readjust to forest life. A number of instances have been recorded of tame elephants wandering off and joining a wild herd. In one particular case, described by the naturalist E. P. Gee, a forest department elephant in Assam went missing for several years and was later recaptured. She still responded to the commands she had been taught and was returned to the care of her mahout. A few months later she delivered a calf.

Wild bulls can sometimes cause trouble with tame elephants when they are being ridden through the forest, though generally the males keep to themselves. As Nazir and I were talking, I asked him about Der Danth, a notorious wild elephant that I remembered from many years ago in Corbett Park. His name means "one-and-a-half tusks" and he had a bad reputation for attacking tourists and park officials. More than likely, Der Danth's broken tusk was the cause of his anger and irritation. As with any tooth, there is a nerve at the core of the tusk, which if damaged can become infected. Rather than being naturally meantempered, Der Danth was probably suffering from a massive toothache.

"He was a *badmash hathi,* a rogue," said Nazir. "The minute Der Danth saw somebody he would chase them. He even charged after cars and jeeps. Der Danth had no fear at all and, in the end, there was no choice. He had to be shot."

Nazir paused to collect his thoughts. "This happened twenty years ago. At the end of the monsoon, when the park is closed to visitors, gangs of laborers are brought in to repair the roads. Sometimes wild elephants harass these people and tear down their tents. But this time, Der Danth killed four laborers. One of them was a woman, eight months pregnant. He picked her up with his trunk and threw her against a tree. The woman died soon afterward but miraculously her child was born on the spot and survived.

"After that incident it was decided that Der Danth had to be destroyed. The park director summoned a hunter from Morad-abad, a famous shikari who had a powerful rifle used for killing elephants. He came here to Dhikala and I was the one who took him into the forest on the back of my elephant. It was not diffi-cult to locate Der Danth and within a few hours we found him near the river. The hunter fired ten shots from his rifle and even then the rogue remained standing. He died on his feet."

Nazir's voice conveyed a note of awe and regret.

"Der Danth's tusks were cut off—the one that was whole and the one that was broken. These were impounded by the for-est department. Then a pit was dug and he was buried. We cov-ered his body with four gunnysacks of salt and four sacks of white lime—to help the remains decompose—which is what we do when our own elephants die."

The following afternoon, when Nazir took us for a ride on Pawanpali, I learned why she is called "tamed by the wind"— because of her speed. Though elephants can run remarkably fast over short distances, they usually sustain a walking pace of eight or ten kilometers an hour. When we set out on Pawanpali's back, she quickly left Mohini and the other elephant behind. Though we cut across rough terrain, through rocky ravines and a marsh choked with dense reeds, nothing seemed to slow her down.

Riding on her back, we saw plenty of wildlife, including a sambar stag who was so complacent that he walked beside us for ten or fifteen minutes. Animals in the forest are often unafraid of a tame elephant with riders on its back, whereas they run away immediately if a human being approaches them on foot. Despite Nazir's best efforts, however, we were unable to see any wild elephants. After circling around to the far bank of the Ramganga we passed by the tiger's kill again. Most of the carcass had been eaten and the only recognizable remains were the cheetal's velvet horns.

On our way back to Dhikala, Nazir took a detour through a marshy area to show us the bones of a wild elephant killed two years earlier in a fight with another male. Bulls often challenge each other over a female in estrus and these battles can be fatal. The skeleton had been picked clean and scattered by scavengers, though most of the larger bones were still intact. Looking down at a huge white femur half buried in the mud, I imagined it could just as easily have been the fossil of a prehistoric elephant, one of the great stegodons that wandered through the Ramganga valley 100,000 years ago. Pawanpali did not seem disturbed by these remains, as she casually waded into a stagnant pool nearby. The dead elephant's vertebrae lay in the water like giant crustaceans. As we moved forward, Nazir nudged me and I looked over my shoulder to see a crocodile slip into the water and glide away between the reeds.

There is no truth to the myth of elephant graveyards, though it is possible that a collection of bones from different animals might be found together near a water hole or a river, because that would be a natural place for an elephant to go if it was sick or injured. The persistent belief that elephants congregate to mourn and bury their dead is a romantic notion based more on human fantasy than zoological facts. As a myth, however, it does suggest the sense of tragedy that we perceive in an elephant's death. Seeing the femur and scattered vertebrae left me with misgivings, as if these were all that remained of the elephant herds that used to roam the grasslands of Corbett Park. The bleached white shapes of the skeleton strewn about in the marsh seemed to signify the extinction of a species that might soon join the fossilized remains of ancient stegodons.

Sighting

The next morning we went for a final drive before leaving the park at noon. All of us were discouraged, especially Negi, who had been promising to show us elephants from the first day we arrived. As he climbed into the car, he told us that someone had seen an elephant the day before, in the forest across the river. Ajay and I were skeptical and Negi did not seem optimistic, but we followed his directions and drove upstream for several kilometers before crossing the Ramganga over a temporary bridge. Aside from the alarm call of a sambar there was little evidence of wildlife. The forest seemed emptier than on any of our earlier drives. Through the dappled colonnade of sal trees the only things we spotted were termite castles and one or two cheetal. The whole expedition seemed fruitless and Negi looked more and more dejected.

By nine o'clock the sun was high above the valley and the mist had long since vanished. The road climbed over the lower ridges of the foothills and eventually descended onto a grassy spit of land that protruded out into the reservoir. Dead trees stood along the water's edge, their leafless branches full of roosting cormorants. Two lines of tire tracks headed straight for the lake but stopped about twenty feet from the shore. We had literally reached the end of the road and there was nothing beyond this point except for the chalky blue water.

Getting down from the car, I looked around without any expectations. Across from us, on the other side of a shallow inlet, were a couple of peahens and two or three deer. Far off in the distance lay the embankment at Dhikala and through my binoculars I was able to spot the roof of the rest house. Scanning the grasslands, I made one last effort to find an elephant, but all that I could see were dozens of gharial crocodiles sunning themselves along the water's edge. These were so far away that even through the binoculars I thought they were logs until one moved its tail.

Just then I heard a whistle and turned to see Negi beckoning. He was facing in the opposite direction, looking up at the foothills.

"There's your elephant," he said, when I got close enough to hear him whisper. "Up there on the ridge."

Raising my binoculars, I focused on the forested slope, and it took a minute before I saw what Negi was pointing at. Five hundred meters away, on the crest of a hill, stood a dull gray shape obscured by branches. At first it looked like a boulder but then it began to move, stretching out its trunk and tearing off a cluster of leaves.

By now Ajay had joined us, and we watched in excited silence as the elephant gradually turned and revealed herself, silhouetted against the sky. A few minutes later another member of the herd appeared nearby and to our delight we spotted a calf standing between the first elephant's legs.

"*Chalo, aap ki yatra safal ho gayi,*" said Negi, under his breath. "At least your quest is now complete."

The elephants ignored us, though they must have sensed that we were there long before we spotted them. After about twenty minutes a young male with stubby tusks appeared on the slope below the other three. Despite the steepness of this hill, the bull scrambled up the ridge, gathering fresh shoots of bamboo in his trunk.

Retrieving my camera from the car, I exposed a couple of frames, though the elephants were much too far away for me to get a clear photograph. The important thing was that we had seen them and I felt immediately relieved. At the end of four days of searching we had finally found wild elephants. Mother and calf stood in profile atop the ridge with perfect serenity, their presence making the forest complete.

on the origin
of elephants

A Message in the Clouds

*T*his story begins with infatuation and desire.

Two lovers separated by a curse, and the poet himself, a slave to the goddess.

Meghaduta (The Cloud Messenger), by Kalidasa.

The poem opens in a lush garden owned by Kubera, the god of wealth. An anonymous yaksha, forest spirit and attendant of the gods, is the caretaker of this garden, but he has a beautiful, sensuous wife and often shirks his duties for the pleasures of her bed. One day, entwined in love, the yaksha forgets to close the garden gate and an elephant wanders in, trampling the sacred blossoms used to worship Shiva. (There is a bee hiding inside a lotus, but that is another story.) For his negligence, the yaksha is banished from the garden and cursed to spend a year alone, away from his wife. He retreats into the distant mountains and is stripped of all his divine powers, living the life of a lonely ascetic. After months of longing and despair, he sees the shape of an elephant in the clouds and asks that it carry a message of adoration to his beloved yakshi.

> *Exiled in the forest, love-starved, emaciated by denial,*
> *The yaksha lets a gold bracelet slip from off his listless arm,*
> *And as the final days of summer draw to an end, he sees a cloud,*
> *Like a musth bull elephant rutting against the mountainside.*

Stirred by unspent passion and struggling to restrain his tears,
Just as the cloud holds back its moisture, this servant of Kubera
Yearns for his wife, knowing that others are returning home
Accompanied by the rain, while he can only dream of her embrace.

For much of *Meghaduta* the heartbroken yaksha instructs the
cloud on the exact route it must follow to reach his wife, describing in lyrical detail the verdant landscape over which the messenger will pass. Having taken the form of an elephant, the cloud
stops along the way to drink from rivers and streams, filling its
trunk with water, then spraying moisture over forests and fields.
The shadowy hue of the storm clouds, which herald the arrival of
the monsoon, is compared to the dark gray color of an elephant's
hide. Being a season of fertility and regeneration, the monsoon
also reflects a bull elephant's cyclical period of sexual arousal,
known as musth. Similar to the rut of a stag, musth is signaled by
excretions from glands on either side of an elephant's forehead.
Though musth can occur at any time of the year, it often afflicts
elephants during June, as the dry summer heat finally gives way to
monsoon rains. During musth, which can last from a few weeks to
several months, the bull becomes temperamental and prone to fits
of anger, as he wanders through the forest in search of a mate.

Thunderclouds engorged with rain, like rampant tuskers—dark
and unpredictable, dripping with the secretions of their temporal
glands. The mountains, too, rise up like elephants, rival bulls
preparing to do battle with the clouds. Bright ivory flashes of
lightning gore the flanks of the ridges as a gathering storm approaches, all curves and angles, sinuous vowels and sharp-edged
consonants—Sanskrit verses marching in procession across a page.

Throughout Hindu mythology, elephants are considered
harbingers of the monsoon because of their love of water. They
are also the vahana, or "sacred vehicle," of Indra, who is Lord of
the Heavens and responsible for the rains. Riding on his majestic

tusker, Airavata, Indra proceeds across the sky with bolts of lightning in his hands, darkening the earth with his shadow and releasing showers of life-giving rain.

The cloud messenger, on its way to find the yakshi, drifts toward Mount Kailash, where Shiva sits in meditation. Most powerful of gods, he cradles the river Ganga in his hair and battles an elephant demon named Gajasura, who has threatened to trample the shrines and huts of mendicants in the forest. After killing Gajasura, Shiva rips off the elephant's hide, which he wears as a bloodstained cloak whenever he performs the tandava dance. Kalidasa recalls this myth in *Meghaduta* with allusions to the clouds that drape the slopes of Mount Kailash, their bruised color similar to Gajasura's skin.

Though an exact date is difficult to determine, Kalidasa lived sometime around A.D. 400. He is often called the Shakespeare of India, author of Sanskrit dramas and narrative verse, romantic bard in the court of Gupta emperor Vikramaditya. Kalidasa seems to have had a particular fascination for elephants, which are mentioned in all of his poems and plays. The metaphor of clouds as tuskers reflects the iconography of the goddess Lakshmi, who sits upon a lotus and is showered by two elephants— an image of fertility. References to elephants as clouds also appear in many other classical Indian texts such as the *Panchatantra,* a collection of allegories and fables. In one of these stories the king of the elephants is described as "a laden cloud with many clinging lightning-flashes. His trumpeting was as deep toned and awe inspiring as the clash of countless thunderbolts from which in the rainy season piercing flashes gleam."

Beyond metaphors and fables, elephants occupy an important place in Sanskrit literature. Gajashastra, or "elephant science," was studied and recorded in several texts that are based on oral traditions going back many years before Kalidasa composed his poems. The *Hastyayurveda,* part of the classic Sanskrit canon, is attributed to the sage Palakapya and describes in detail the methods of capture, care, and training of elephants. Kautilya's *Arthashastra,* a treatise on statecraft, contains chapters that provide instructions for the superintendent of elephants, an important post in ancient empires.

A court poet like Kalidasa would undoubtedly have been aware of the lore and vocabulary of Gajashastra. The habits and idiosyncrasies of elephants must have been a regular subject of conversation in the court. For that reason, his image of an elephant in the clouds is more than just a mythological motif. In fact, Kalidasa's metaphor in *Meghaduta* is clearly based on the observed behavior of elephants.

When he describes the cloud butting against the mountains, like a tusker ramming into a riverbank, the poet accurately portrays a bull elephant in musth. Scientists who have studied elephants in the wild often report that adult males will seek out a water hole or riverbank and use their tusks to rut in the mud. As the naturalist M. Krishnan writes in the journal of the Bombay Natural History Society:

> A peculiarity noticed in tuskers in musth is that they often carry tight-packed clay on their tusks. . . . This adherent clay is acquired when the tusker gores earth banks and even the clayey bottoms of forest pools while in musth: this goring is not something done in a frenzy, but evidently indulged in to cause by the pressure imposed on the swollen glands the free outflow of musth from the temporal glands.

The fluid that streams down the face of a musth elephant is often described by Sanskrit poets as the supreme elixir of love and passion. In the *Ramayana* epic, when Rama and Sita are married, the streets of Ayodhya are flooded with the temporal juices of royal elephants. Honeybees are said to swarm around an elephant in musth, attracted by the cloying sweetness of its secretions. And the monsoon rains, which bring the fecund earth to life, are repeatedly compared to the flow of musth. Many translators have substituted the term *ichor* for musth, which is not entirely accurate but conjures up an interesting comparison. In Greek mythology ichor is the liquid that circulates through the veins of the gods, though in a secondary and less appealing sense, it also denotes pus that issues from a boil.

One of the intriguing aspects of Sanskrit literature is the way in which poetry and science are often inextricably linked. In the *Matangalila,* a rendition of Gajashastra composed by the Sanskrit poet Nilakantha, he describes an elephant in musth:

> With honey-colored nails, tusks, and eyes, skin like a dark cloud, red corners of the eyes, lotus-filament spots (on the skin), quarreling with other elephants, with sporting in dust and water the handsome elephant king becomes "temple-filled" (in the first stage of *must*). His cheeks are washed with the *Must*-fluid that flows in streams; he is filled with thunder (roaring) like a rolled-up cloud; rushing forward intent on slaying even those at a distance.

Later, Nilakantha compares an elephant in the final stage of musth to "a cloud that has discharged its accumulation of water."

Reading these passages alongside *Meghaduta* it is obvious that Kalidasa was familiar with elephant lore and science. His opening metaphor, taken straight from Gajashastra, becomes all the more potent and provocative when we connect the lovelorn yaksha's longing for his wife with a musth elephant's state of unrequited passion. In the poet's mind—where everything is possible and nothing is denied—the cloud, the elephant, and the lover become one.

Myth and Science

When the cosmic egg was first cracked open by the Creator, its yolk became the sun and the other substances it contained became the oceans, the land, and the sky. According to the *Matan-galila,* the two halves of this golden eggshell were then presented to a chorus of divine sages, whose chanting accompanied the formation of the world. From out of the eggshell in the Creator's right hand emerged eight male tuskers, including the magnificent Airavata who would carry Indra, Lord of the Universe, on his back. From the eggshell in the Creator's left hand emerged eight female elephants. This original herd of sixteen were the progenitors of the species and their offspring "ranged at will over the forests, rivers, and mountains of the whole world."

In the earliest of times, elephants were able to change shape and size, like clouds in the sky. They could take to the air or swim in the sea, and they wandered unimpeded over the four quarters of the earth. During this mythical age of purity and peace, an enormous banyan tree grew at the foot of the Himalayas. Under its spreading branches lived an ascetic named Dirghatapas, who spent long hours in meditation. Like most sages he was a saintly but short-tempered man. One day a herd of elephants came flying overhead and alighted on the banyan tree, breaking a branch that fell onto the hermitage of Dirghatapas. Furious at having his contemplation disturbed, the sage cursed the herd of elephants, saying that they would no longer be able to fly about as they wished. Instead, man would take away their freedom and use them as beasts of burden.

In the nearby kingdom of Anga ruled a king named Romapada. His capital was the city of Champa on the banks of the river Ganga. While Romapada was giving audience to a conclave of mystics and philosophers, reports arrived that elephants were raiding crops and harassing the farmers in his kingdom. Hearing this, the mystics granted the king a boon, allowing him to catch and tame the elephants.

Immediately, Romapada sent a party of men to hunt down and capture the herd. They found the marauding elephants grazing in the jungle near the hermitage where another sage,

Samagayana, and his son Palakapya lived. Trapping the elephants, the hunters brought them back to Champa, where they were securely tethered to posts in the ground. Soon afterward, however, Palakapya arrived in the capital. He went directly to the captive elephants, greeted them affectionately, and healed their wounds with soothing herbs and ointments.

Seeing this, the king's attendants asked Palakapya why he was treating the elephants with such compassion, but the sage gave no reply, for he had taken a vow of silence. Eventually Romapada himself came to see what was happening. The king approached Palakapya with deferential respect and even washed his feet. Still there was no explanation for the sage's behavior and only when the king bowed down in front of Palakapya did the hermit finally begin to speak. At this point he told the story of the origin of elephants, how they had emerged from the cosmic eggshell, how they had once roamed wherever they pleased, and how one of the herds had been cursed by Dirghatapas.

After revealing all of this, Palakapya then told the story of his own birth. He explained that his mother was a yakshi, or "nymph," named Ruchira, whom the Creator had endowed with every aspect of beauty. Unfortunately, her physical charms distracted yet another irritable ascetic from his meditation and he turned her into an elephant. The only way Ruchira could release herself from this curse was to conceive a son. After wandering through the forest in desperation, she arrived at Samagayana's hermitage one morning. The sage had just woken up from a dream in which he had seen a vision of a seductive yakshi. When he went outside to relieve himself, some of his sperm mixed with his urine and the elephant rushed forward and drank it up.

Soon afterward a boy was born from Ruchira's mouth and she was restored to her original form. Before ascending to heaven, however, the yakshi left her child in the care of Samagayana. The boy was named Palakapya and he grew up among the elephants, "roaming with them through rivers and torrents, on mountain tops and in pools of water, and on pleasant spots of ground, living as a hermit on leaves and water, through years numbering twice six thousand, learning all about the elephants, what they should and should not eat, their joys and griefs, their gestures and what is good and bad for them."

Most of the *Matangalila* consists of a dialogue between Palakapya and Romapada, in which the hermit explains to the king about the proper care and treatment of elephants, precise methods for their capture and training, as well as the different traits and qualities they exhibit. Sections of the text seem to have been copied or paraphrased from the *Hastyayurveda,* of which Palakapya is the author.

Along with a detailed mythology about the origin of elephants, the *Matangalila* is filled with observations regarding their biology and behavior. Unlike most modern texts, which separate poetry from prose, Nilakantha's work makes no distinction between imagination and reason. In fact, for him the fantastic and the scientific seem to reinforce each other, and the commentary he provides on elephants, related through the words of the sage Palakapya, is rich in metaphor as well as empirical facts.

In his discourse with the king, Palakapya goes into great detail regarding the anatomy of elephants. Most Sanskrit texts contain a considerable amount of categorization and the *Matangalila* is no exception. Palakapya provides lists upon lists, identifying the various parts of an elephant, from the gandusa, or "tip of the trunk," to the root of the tail, called the pechaka. He catalogues their personalities and temperaments, using categories generally ascribed to men, such as the humors of rajas or tamas, which denote "a fiery hot temper" or "a dark and brooding melancholy." Palakapya also divides elephants into three castes: the bhadra, or noble tuskers suitable for carrying a king; the manda, who are slow and stolid, an ordinary kind of elephant; and the mriga, who are lean and long-legged, capable of swift movements like a deer.

More than the encyclopedic compilation of traits, however, the most striking aspect of the *Matangalila* is the way in which the myths themselves reflect an intimate knowledge of the elephant's physiology and habits. Clearly, the sages and mystics of ancient India, who sought refuge in the forest, must have come in contact with wild elephants on a regular basis and had firsthand experience of their behavior. Unlike Palakapya, who shared a familial bond with these animals, many ascetics found elephants to be destructive and disturbing. In one particular legend, concerning the Vedic sage Bhrigu, a roaming herd of elephants entered his hermitage and ravaged the surrounding orchards, breaking branches and eating the fruit. They also trampled the sage's hut and extinguished his sacred fire by urinating and defecating on the hearth. When Bhrigu saw the damage they had caused, he was furious and cursed the elephants, saying that henceforward they would feed on their own dung and urine.

Aeons later, the zoologist Raman Sukumar in his book, *Elephant Days and Nights,* describes observing young elephants tasting the dung of adults in a herd. He explains this behavior, known as coprophagy, as a method by which the juvenile elephants are able to ingest microbes that produce enzymes required for digestion. Since these enzymes do not occur naturally in an elephant's stomach and intestines, they are passed on from one generation to the next through the curse of Bhrigu. Once again, the scientific evidence and the mythological narrative converge.

Survivors

Out of dozens of ancestral variations, only two species of the proboscidean order are still found today: *Elephas maximus* in Asia and *Loxodonta africana* in Africa. A subspecies of African elephant, *Loxodonta cyclotis,* has been identified, but essentially these are the last two members of the *Elephantidae* family that remain on earth, following the extinction of the woolly mammoth at the end of the last ice age. Though similar in many aspects, Asian and African elephants have several anatomical differences.

On average, *Loxodonta* stands a foot and a half taller than *Elephas.* Some males in Africa have been measured over twelve feet at the shoulder, whereas the biggest elephants in India are no more than ten feet tall, and often quite a bit smaller. The relative weight difference between the two species is also considerable. An African bull can weigh over six tons, while Asian males weigh closer to four tons. The ears on an African elephant are broader and more dramatic, flaring out like leathery sails, whereas the Asian elephant has smaller, more delicate ears that fold over at the top and become pleated with age. The shape of an Asian elephant's ear is sometimes compared to the map of India, being wider at the top and tapering down to a point.

While the skin of both species is similar, the African has a rougher hide and the trunk in particular is more wrinkled. Asian elephants often have a mottled pink complexion around their face, which grows lighter with age. The shape of the skull is also different, with *Loxodonta* having a more extended and tapered cranium, while *Elephas* has a flatter face, with a bulging forehead. Though both species have similar skeletal structures, the Asian elephant is humpbacked, whereas the African is swaybacked.

Less prominent differences have been identified, including striations on the molars, which give *Loxodonta* its name, as they are "lozenge-shaped." *Elephas* has teeth that are ribbed on top, like a coarse-cut file. The tip of the trunk on an African elephant is equipped with two prehensile fingers that are used for feeling and plucking; the Asian elephant has only one finger that curls over its nostrils like a protective hood. Toenails have often been pointed out as a distinguishing feature, with the African elephant having five on its forefeet and four on its hind feet, while the

Asian elephant often has five on each. This distinction, however, is misleading as some forest elephants in Africa have been found with five toenails on all four feet and the exact number on Asian elephants is unpredictable, though twenty is considered an auspicious number.

Comparisons between the two species have generated a number of misconceptions, often perpetrated by colonial writers, whose powers of observation were colored by racial and cultural stereotypes, including Victorian notions of physiognomy. The African elephant has been unfairly portrayed as a more primitive and aggressive creature that cannot be tamed. Asian elephants, on the other hand, are often described as more intelligent and compatible with man. The truth is that both species can be tamed, but in India there has been a much longer tradition of capturing and training elephants. On the other hand, ancient Greek and Roman armies used both *Loxodonta* and *Elephas* as war machines. In the Belgian Congo, during the early part of the twentieth century, successful attempts were made to train African elephants to work in logging operations, though these were never consistently put into practice. More recently, African elephants at a safari park in Botswana have been trained to give rides to tourists.

For many years it was believed that the two species could not be crossed, as all attempts to interbreed an African and an Asian elephant were unsuccessful. However, in 1978, at the Chester Zoo in England, an Asian female named Sheba mated with an African male named Jumbolino. Against all predictions Sheba became pregnant and delivered a premature male calf on July 11, 1978. The baby elephant had only one finger on its trunk and the same number of toenails as its mother, but its ears and the shape of its head were recognizably those of an African elephant. Sadly, the calf lived for only two weeks, proving that conception between the two species was possible but leaving open the question of survival. The remains of this crossbred elephant—stuffed and mounted by a taxidermist—are preserved in London's Natural History Museum.

Taxonomy and Extinction

The architecture is reminiscent of Christendom's greatest sanctuaries, with towering stone spires, arched windows, and ornate lintels. Vaulted chambers rise up to high ceilings that evoke a sense of infinite awe. Stained glass windows add to an atmosphere of ethereal beauty and magnificence, while the staircases to upper galleries are like the tiered approaches to heaven. Though the building itself conveys an aura of religious grandeur, it stands for the antithesis of faith. Instead of being dedicated to the worship and adoration of a divine creator, this cathedral of science was intended to celebrate the doctrines of natural selection.

The grand entrance hall of the Natural History Museum in London is dominated by the bones of dinosaurs rather than the symbols and relics of religious heritage. Replacing the statues of saints are the marble busts of Darwin, Huxley, and Owen. Yet the inescapable impression remains that this venerable museum and all that it contains share many of the same traditions as the church. Whether it is the Latin names that identify each of the specimens or the way in which the towering ceilings evoke eternity, one realizes that the denial of a creator does not preclude the mysteries of creation. Like any orthodoxy, Darwinian theories have assumed an enduring structure, and the Natural History Museum, as much as the bones, skin, and feathers that it preserves, represents a cornerstone of that legacy.

The museum's large gallery of mammals contains stuffed specimens of Asian and African elephants, though these are dwarfed by an enormous, life-sized model of a blue whale, suspended from the ceiling alongside its skeleton. This juxtaposition of the world's largest mammals, both on land and in the sea, suggests a primal link. There is ample evidence that the earliest ancestors of *Elephas maximus* once lived in the ocean. Anyone who has watched an elephant bathing knows that they are excellent swimmers, who revel in the water as if it were their natural element. In the Andaman archipelago, off the eastern coast of India, there are elephants that swim from one island to the next, crossing as much as thirty kilometers of open sea. These animals were originally introduced from the mainland for logging operations. Later, after being released into the wild, they formed a feral herd. The Andaman elephants have adapted to their island environment and seem to have no trouble swimming in salt water. Several documentary films have been made of these elephants with underwater cameras. Watching them move through the turquoise waves, one can easily appreciate their aquatic origins.

Furthermore, the closest living cousin of the elephant is the sea cow—both the manatee of North and South America and the dugong, which makes its home in the Indian Ocean, where it is found in the coastal waters of the subcontinent as well as on islands like the Andamans. Scientists have linked these species to the elephant because of common physical traits, primarily the location of their genitalia. The vulva on a female elephant is close to the middle of the abdomen, as it is on a sea cow and other marine mammals. Similarly, the testes of a male elephant and the male dugong are internal. The bone structure of the sea cow also bears similarities to the elephant, particularly its vestigial feet. In addition, the dugong has small tusks and a bulbous nose. There has been some speculation that the elephant's trunk evolved as a kind of snorkel, to help it breathe while submerged in water, though this remains open to debate, as the trunk serves so many other purposes on land.

Indian mythology provides a number of legends and images that suggest the aquatic origins of the elephant. By some accounts Indra's tusker, Airavata, emerged from the churning of the primordial sea. Several ancient carvings exist of fish-tailed elephants, particularly a first-century B.C. sandstone medallion from Mathura that looks very much like a mermaid, except with the head of an elephant. A more threatening visage can be found in the image of a sea monster, known as the Magor, which is often depicted as a crocodile with a distinctly elephantine trunk.

Arranged along one wall of the museum's gallery of mammals is a display titled FORERUNNERS OF THE ELEPHANT. The first model, scaled down to one-eighth of its original size, is of *Moeritherium,* a piglike ancestor with a blunt snout and sharp incisors that lived forty million years ago. Next to this stands an image of *Dinotherium,* which has a recognizable trunk but whose tusks protrude downward from the lower jaw like ivory plowshares. Other orthodontic incarnations follow until the more recognizable *Cuvieronius* appears, based on a specimen found in Argentina. This animal has a longer, less graceful body than today's elephants, though the tusks and trunk are similarly positioned.

Elephants provide some of the most dramatic examples of evolutionary change. Apart from their size, the two distinguishing features readily apparent to the untrained eye are their trunks and tusks. Zoologists may focus on more obscure traits, such as an elephant's molars and the location of the genitals and mammary glands, but the most obvious aspects remain in full view. Particularly intriguing, when one traces the family tree of the proboscidean order, are the remarkable variations that occur in the shape and size of the tusks and trunk. It has been suggested that the reason for such dramatic changes lies in the widespread migration of prehistoric elephants and their need to adapt to different climates and terrain. This reminds us of Nilakantha's claim that primordial elephants "could go anywhere they pleased, and assume any shape." Viewing an exhibit on the evolution of the elephant is like looking at a sequence of caricatures in which the recognizable features of this animal have been exaggerated and juxtaposed in unpredictable and seemingly fanciful variations.

The Natural History Museum also contains the skull of a *Stegodon ganesa,* one of at least eight different species of proboscideans that lived in India before the advent of man. Its enormous tusks curve outward and then converge like giant calipers. Next to this stand more recent variants like the mammoth, which appears in cave paintings from the Stone Age. Neanderthal tribes in Europe hunted these to extinction. In a glass case nearby is a specimen of a mammoth's fur, from frozen remains unearthed in Siberia. Its hair looks like dried coconut husks. The skull is also exhibited, complete with two magnificent tusks that splay out in opposite directions.

Like most elephants, the mammoth had a sinus cavity in the center of its head that looks like a large eye. Several writers have speculated that the Greek myth of the Cyclops is based on a mammoth's skull. For someone unfamiliar with an elephant, the monstrous cranium with a nasal socket in the forehead and protruding tusks would certainly conjure up images of a one-eyed giant.

The exhibit on elephants concludes with a display that shows the molars of *Elephas maximus,* which are much more crucial to its survival than its tusks. These teeth, which protrude from the inner jaw, are about six to eight inches long and four inches wide. Ridged on top like a washboard, they are used to shear the two to three hundred kilos of grass and leaves that an elephant consumes each day. Because of its longevity and the vital role these molars play, nature provides elephants with six sets of teeth during their lifetime. As each of these is worn down from constant chewing, another line of molars pushes in from behind. Once the sixth set of teeth has been ground smooth, usually by the age of seventy, the elephant starves to death.

Vanishing Ivory

Unlike in Africa, where both male and female elephants bear tusks, in Asia only males are tuskers. Females of the species have small incisors hidden below their upper lips, which are visible when they raise their trunks. In addition, approximately 40 percent of Asian male elephants do not have tusks and are known as makhnas. Some regions of the subcontinent, such as Assam, have many more tuskless males than tusked. Sri Lanka has the lowest proportion of tuskers in all of Asia, about 10 percent.

The phenomenon of tuskless males has been a source of considerable comment and debate. Makhnas are identical to other bull elephants, except for their lack of ivory. Some colonial writers considered them larger and more dangerous, though there seems to be little more than anecdotal evidence to support this claim. The absence of tusks, however, does not stop a makhna from going into musth. Even without his ivories, a bull elephant can do a great deal of damage, though in combat with other males he suffers a disadvantage. Tusks are used as weapons when rival bulls square off for the favors of a cow. Nevertheless, a makhna is equally able to mate with females in a herd and propagate his own bloodline.

Theories have been put forward that there are more makhnas in places like Assam because of the value that human beings have placed on ivory, whether as a trophy or as a commodity. It would seem logical that if males with tusks were the only animals targeted by hunters, then tuskless bulls would multiply more quickly. Colonial sportsmen, steeped in the obsessive and often misguided traditions of British fair play, set down a code of regulations in the forests of Assam in a futile attempt to even things out. Hunting licenses issued under the provisions of the Elephant Preservation Act of 1879 specified that the tusks of any elephant killed in Assam were to be impounded by the forest department and could only be reclaimed as trophies after the hunter produced evidence of shooting a makhna as well.

Scientists who have studied tuskless males confirm a direct correlation between the population of makhnas in a particular forest and the pressures of elephant poaching. For instance, in

Corbett National Park, where this has only recently become a problem, there is a relatively healthy ratio of one tusker to every four females, whereas in Periyar Tiger Reserve in Kerala, the ratio is one to sixty. The contrast is almost as dramatic between elephants in Karnataka's Nagarhole sanctuary, where tuskers are plentiful and the ratio is approximately one to five. Just across the Kabini River, however, in the adjacent forests of Bandipur and Mudumalai, the ratio drops dramatically. Even a nonscientist can easily deduce that the areas where makhnas and females greatly outnumber tusked males correspond to the regions where gangs of poachers have been operating with impunity. Raman Sukumar sums up the problem:

> If tuskers continue to be killed at the present rate there is a danger that there will be too few of them left to breed, resulting in a decline in fertility. The tuskless males will, of course, be at an advantage and will contribute their genes in greater proportion than will the tuskers to future generations. If such a scenario were to unfold, the makhnas will come to gradually dominate the population. This might take several elephant generations, given the slow rate at which elephants multiply and die. The elephant itself might continue to survive as a species . . . but then, to many, the beauty of an elephant is in its tusks.

Most scientists are cautious when it comes to identifying contemporary examples of natural selection, believing that lasting changes in a species occur over thousands of years rather than a couple of centuries. There is also the possibility—which cannot be discounted until genetic testing has been done—that female elephants may carry the gene responsible for tusks. Nevertheless, it would seem that *Elephas maximus* stands at the brink of one of the most dramatic changes in its physiology. Though it may take centuries to imprint this change in its genes, judging from scientific and anecdotal evidence, the Indian elephant is in danger of losing its tusks. Of course, there is always the greater risk of extinction, which will erase these animals altogether, but to a certain extent their survival as a species may depend on how quickly they jettison their ivory.

Charles Darwin wrote surprisingly little about elephants, though they represent some of the most convincing examples in support of his theories. *On the Origin of Species* refers to the elephant only twice, once to illustrate the adaptability of animals to varying climates (i.e., woolly mammoths surviving in the Ice Age and today's elephants living in tropical forests), and once to calculate the rate of increase in population.

> The elephant is reckoned the slowest breeder of all known animals and I have taken some pains to estimate its probable minimum rate of natural increase; it will be safest to assume that it begins breeding when thirty years old, and goes on breeding till ninety years old, bringing forth six young in the interval, and surviving till one hundred years old; if this be so, after a period of from 740 to 750 years there would be nineteen million elephants alive, descended from the first pair.

Setting aside the obvious inaccuracies (elephants begin to breed well before thirty and do not live to a hundred), Darwin offers an unusually optimistic calculation of the elephant's ability to procreate. If only it were true.

The more intriguing references to elephants in Darwin's work come from his book *The Expression of the Emotions in Man and Animals.* In one of the chapters he writes, "The Indian elephant is known sometimes to weep." He quotes Sir Emerson Tennent, a colonial official in Ceylon (now Sri Lanka), who reports recently captured elephants expressing despair through "tears which suffused their eyes and flowed incessantly." Another elephant taken from the wild was similarly described. "When overpowered and made fast, his grief was most affecting; his violence sank to utter prostration, and he lay on the ground uttering choking cries, with tears trickling down his cheeks."

In much of the chapter about weeping, Darwin focuses on the contraction of facial muscles near the eye, the *orbicularis oculi.* He argues that the tightening of these muscles when a human being is angry, frightened, or upset represents a unique characteristic in man and is related to the shedding of tears. Experiments were conducted at the zoological gardens in London, where keepers ordered their captive elephants to trumpet and scream. Darwin observes that in the case of Indian elephants the orbicular muscles clearly contracted, though no tears were shed. As for African elephants in the zoo, he reports that their facial muscles did not contract.

It seems unusual that Darwin focused on this particular aspect of the elephant rather than the more obvious physical characteristics, such as its trunk and tusks, of which he makes no mention. Perhaps this was because the father of modern biology never traveled to India and only observed elephants briefly in a zoo. Whatever evidence Darwin gathered about their behavior and biology seems to have been largely anecdotal. Though one can only speculate, it is very possible that if he had enjoyed the opportunity of studying these creatures in the wild, Darwin might have noticed that in addition to shedding tears they were also shedding their tusks.

Dasharatha's Arrow

The *Matangalila* contains no mention of hunting elephants for their ivory. Instead, tusks are considered an integral aspect of the animal's grandeur and vitality. The most beautiful tusks are described as "honey colored" and those who attend upon an elephant are instructed to anoint and polish them with care. Nilakantha explains that when an elephant is sick or in a weak-ened state, its tusks become loose and sometimes fall out. Clearly, their value lies not in the ivory itself, separated from the elephant, but in the way the tusks adorn a living creature, signifying its strength and majesty.

In most Sanskrit texts elephants are regarded as symbols of royalty and the removal of their tusks is equated with evil and immoral forces. The *Ramayana* epic contains stories of the demon king Ravana who gains terrible and malicious powers through long years of meditation and austerity. He acquires such horrific strength that he is able to shake the foundations of Mount Kailash and defeats Indra in battle. During this conflict, Ravana fights against the primordial herd of elephants who were born from the cosmic egg and support the four quarters of the cosmos. In his fury, he grapples with the elephants and tears out their tusks by the roots. Years later, when Ravana is finally defeated by Rama, an old scar is discovered on the demon's body, where one of the elephants had gored him with its tusk.

Much of the tragedy and turmoil of the *Ramayana,* beginning with the exile of Rama and Sita to the forest and ending in the prolonged battle between heroes and demons, can be traced back to a hunting accident in which a man is mistaken for an elephant. Dasharatha, king of Ayodhya and father of Rama, was an expert archer who prided himself on the accuracy of his aim. One night he was alone in the forest, testing his skills by hunting in the dark. The king needed only to hear the slightest sound an animal made and he could send his arrow to its mark.

On that moonless night, Dasharatha positioned himself near the bank of a river where he waited for his prey to come and drink. After awhile he heard a gurgling sound and assumed it was an elephant sucking water into its trunk. Drawing his bow, he shot an arrow through the darkness, but instead of the trumpeting of a wounded elephant he heard an anguished human cry. Rushing down to the river, the king discovered that he had shot a young ascetic. The sound he had mistaken for the elephant drinking was the gurgling of a water vessel that the ascetic had come to fill. Mortally wounded and in great pain, the young man told Dasharatha that he lived in a hut nearby with his aged parents, both of whom were blind. He pleaded with Dasharatha to go and tell them what had happened, for they were alone and would be anxious because he hadn't returned. The ascetic also begged the king to pull the arrow from his chest and let him die. With great reluctance, Dasharatha complied, then refilling the water vessel, he carried it to the ascetic's hut where the blind couple was waiting. Hearing his footsteps, they assumed it was their son, but after giving them water to drink, the king revealed his identity and explained what he had done. The parents were distraught and asked to be taken to the riverbank where preparations were made for a cremation. Having no one else to look after them and overcome with despair, the elderly couple threw themselves onto the burning pyre. Before the flames consumed them, however, Dasharatha was warned that because of his actions he, too, would die of grief over the loss of his son.

Many years later this curse came true, when Rama left Ayodhya for the forest, after which the king collapsed in sorrow and breathed his last. Through a trajectory of cause and effect, the destiny of all the major characters in the *Ramayana* can be traced back to Dasharatha's arrow, which propels a tragic chain of events and raises issues of guilt, accountability, duty, and fate. The ascetic's death was an accident, for which the king bears responsibility, but his true intention was to kill an elephant, which also affects his fate as it is a symbol of royalty. At the core of this myth lies an essential confusion between elephant and man, compounded by the device of the gurgling water vessel. Unlike the elderly parents, whose loss of sight was a symptom of age and infirmity, Dasharatha's blindness was entirely of his own making and reflects his arrogance as a hunter and marksman. The king's decision to shoot an arrow at what he believed was an elephant ultimately brought a curse upon his kingdom and led to his own demise.

Anatomy Lessons

The well-known parable of the blind men and the elephant is another telling story of misperception. Each of the men touches a different part of the animal's body and describes it according to his limited powers of observation. The blind man who encounters the trunk identifies it as a serpent; the next, who touches its leg, decides that this must be a tree; the third, who lays a hand against its flank, is convinced that it can only be a wall; the fourth feels its ear and insists it is a fan; the fifth man, who grabs the animal's tail, believes it is a rope. None of the blind men is able to conceive of the elephant as a whole.

Many early descriptions of elephants by Europeans were equally misinformed. For instance, when the bones and tusks of prehistoric mammoths were first unearthed in Estonia, it was believed that these were the remains of a giant species of rat that lived underground. Confronted for the first time by war elephants, Roman soldiers described them as huge dragons with snakes on their heads. Early pictorial representations of the elephant in medieval bestiaries often portrayed their tusks jutting upward like the horns of a cow or pointing in other curious directions, while their trunks flared out like trombones.

When it comes to names assigned to elephants, the etymology also reveals conflicting perspectives. Gaja, the Sanskrit word for "elephant," shares the same root as garj, or "thunder," which underscores the metaphorical connection between elephants, storm clouds, and the monsoon. The modern Hindi word hathi comes from the Sanskrit hastha or hasthin, which means "hand," alluding to the animal's versatile trunk. In contrast, the Greek word elephas, from which both the Latin and English names are derived, means "ivory." The emphasis placed on the trunk in Indian vocabulary implies an appreciation and understanding of the animal's dexterity as well as its usefulness to man. On the other hand, defining an elephant by its ivory would suggest differing priorities.

Many of the earliest Greek and Roman writers make reference to the ivory trade and discuss the means and methods of hunting elephants. In A.D. 77, Pliny the Elder described the elephant in his *Naturalis Historia:* "The largest land animal is the elephant, and it is the nearest to man in intelligence: it understands the language of its country and obeys orders, remembers duties that it has been taught, is pleased by affection and by marks of honour, nay more it possess virtues rare even in man, honesty, wisdom, justice, also respect for the stars and reverence for the sun and moon."

Despite these accolades, Pliny makes it clear that many of his contemporaries considered tusks to be the primary object of value in an elephant. He contends that the animals also recognize this fact. "They themselves know that the only thing in them that makes desirable plunder is in their weapons which Juba calls 'horns' but which the author so greatly his senior, Herodotus, and common usage better term 'tusks'; consequently when these fall off owing to some accident or to age they bury them in the ground."

Pliny goes on to explain how elephants have an instinctive fear of man because human beings kill them for their ivory. He describes how they "tremble in fear of an ambush" and warn each other about the presence of hunters in the forest. "Why should they dread even the sight of a man himself when they excel him so greatly in strength, size and speed? Doubtless it is Nature's law and shows her power, that the fiercest and largest wild beasts may have never seen a thing that they ought to fear and yet understand immediately when they have to fear it." Almost two millennia after Pliny wrote these words, they carry an ominously contemporary resonance, particularly when he adds that "an ample supply of tusks is now rarely obtained except from India, all the rest in our world having succumbed to luxury."

During the second century A.D. an encyclopedia of animals, *Physiologus,* was published in Alexandria and later translated and distributed throughout the Mediterranean world. This early bestiary is full of exaggerations, many of which are intended to convey Christian themes and messages. Elephants, for example, are described as animals "whose copulating is free from wicked desire." The anonymous author of *Physiologus* also reports that elephants give birth only in water and that they have no knees, their legs being straight and unbending. This peculiarity means that when they fall down they cannot get up; hence elephants must sleep while leaning against trees. Hunters are instructed to weaken a tree by cutting halfway through its trunk, so that when an elephant rests its weight against the tree, both will fall and the animal is rendered helpless.

Many of these misconceptions tell us as much about those who provide erroneous information as they do about the elephant itself. Greek and Roman soldiers who encountered tuskers on the battlefield would undoubtedly see them as larger and more terrifying than they really were, with armored skin and lancelike tusks. Both ancient and modern European writers who traveled to India often sought to promote a sense of awe and wonder in the minds of their readers by portraying the elephant as an exotic and mysterious beast, exaggerating stories about its intelligence as well as its physical attributes. Being as large as it is, the elephant inspires hyperbole and embellishment.

Twice round an elephant's foot is a true measure of its height.

This simple formula is remarkably accurate, though even today most accounts of an elephant's size are prone to stretch the truth. Despite reports of Indian tuskers eleven or twelve feet tall, experts agree that ten feet is the maximum height for *Elephas maximus.* Adult males usually stand between eight to nine-and-a-half feet at the shoulder and cows are seldom more than eight feet tall.

Much of the blame for these inflated dimensions lies with colonial hunters who often measured a dead animal "over the curves," which adds more than a few inches. Weighing an elephant—whether dead or alive—is even more difficult than gauging its height, and most of the time, except when proper equipment is available, it is purely a matter of estimate and speculation. The other aspect of an elephant's anatomy that interested hunters was the exact size and location of its brain. Their motives had nothing to do with scientific inquiry but with where to place a fatal bullet.

As Darwin's comments demonstrate, the longevity of elephants was also overestimated. Early Greek writers claimed that the elephant had a life span of two hundred years and until recently most people believed that elephants lived well beyond a hundred. One reason for this may have been that the life expectancy of human beings used to be far less than it is today and many tame elephants outlived their masters. Now it is generally accepted that seventy years is the maximum age for an elephant, with rare exceptions.

Early British efforts to study the physiology of elephants in a scientific manner range from Patrick Blair's "The Anatomy and Osteology of an Elephant, Being an Exact Description of All the Bones of the Elephant Which Died near Dundee, April the 27th, 1706, with Their Several Dimensions," to Richard Owen's "Description of the Foetal Membranes and Placenta of the Elephant (*Elephas Indicus,* Cuv) with Remarks on the Value of Placentary Characters in the Classification of the Mammalia." These and other sources are cited in a slender but comprehensive book, *Anatomy of the Indian Elephant,* by L. C. Miall and F. Greenwood, who dissected a young female elephant in 1874. This specimen was purchased, probably from a circus, by the Council of the

Leeds Philosophical and Literary Society, who supported their research.

Miall and Greenwood quickly discredit the idea that an elephant has no knees by providing thirteen pages of description on the foreleg alone, along with a complete diagram of its musculature. Their writing suffers from Victorian verbosity but the fastidious details of their research are impressive. One aspect of an elephant's anatomy they focus on are the reproductive organs. Here the authors begin by quoting a passage from Aristotle's *Historia Animalium,* which provides a surprisingly accurate description: "The penis of the elephant is like that of the horse, but small considering the animal's bulk. The testes are not visible externally, but are placed inside near the kidneys. The pudendum of the female is placed in the position which in ewes is occupied by the dugs, but for congress is drawn upwards, so as to facilitate the action of the male. It is naturally a wide orifice." Miall and Greenwood note that the clitoris on their specimen measured fourteen inches.

Sexual intercourse between elephants has long fascinated human observers, partly because it is so rarely witnessed. In particular, the location of the female genitals led to all kinds of speculation, including a belief that elephants were only capable of copulating in water because of their weight. Muhammad al-Damiri's Arabic bestiary, *Hayat al-Hayawan,* promotes this theory, and he goes on to quote a source: "Al-Kazwini states that the vulva of the female is situated in (under) the groins (armpit), and that when it is the time for covering, it rises up and comes forth for the male, so that it then becomes possible for the male to have coition with her. Celebrated be the praises of Him to whom nothing is impossible!"

Four centuries later, in 1755, Samuel Johnson's dictionary included the following entry on elephants: "In copulation the female receives the male lying upon her back; and such is his pudicity, that he never covers the female so long as anyone appears in sight." An 1803 lithograph by the French naturalist J. P. L. L. Houel "whose authority is not unimpeachable," according to Miall and Greenwood, shows two elephants mating on solid ground, in what might be described as a missionary position. Miall and Greenwood preferred to accept a later account by Dr. Morrison Watson, who "informs us that the evidence of eyewitnesses, though not so ample or explicit as might be desired, goes to show that the female rests upon the fore-knees with the hind legs extended in the standing posture." Once again the bending of knees comes into play.

A much more recent and reliable account of elephant copulation can be found in Raman Sukumar's *Elephant Days and Nights,* in which he explains that the location of the female genitals does not pose a problem for the male since "the fully erect penis is flexed upwards near the tip so that it may easily hook on to the vaginal opening." Sukumar's field notes, in which he records observing elephants mating, describe the bull mounting the cow from the rear, "with his hind legs half crouched and front legs astride." The duration of the act he witnessed was roughly thirty seconds and it was repeated several times over the space of a couple of hours.

Sukumar also explains the manner in which female elephants send out "chemical cues" to signal when they are in estrus. These pheromones are present in the female's urine, which a bull elephant samples with the tip of his trunk. In a process known as flehmen, the male then puts the end of his trunk against a vomeronasal gland in the roof of his mouth, which indicates if the cow is in estrus or not. These same pheromones are also found in the cow's vagina and bull elephants often "check" a female by touching her genitals with their trunks.

Further insights into this subject can be found in D. Mariappa's *Anatomy and Histology of the Indian Elephant,* published in 1986. The author is a former professor of anatomy and associate dean at Madras Veterinary College. He conducted his research on three elephant fetuses, one of which appears in a photograph as the frontispiece for his book. Dr. Mariappa is as fastidiously scientific as his Victorian predecessors and goes into even greater detail when describing muscles in an elephant's knee, such as the *Extensor carpi radialis* and *Flexor communis digitorum*.

One of Mariappa's most important discoveries, however, is neurological and focuses on the tip of an elephant's trunk.

> A transection of the prehensile finger shows the skin forming a covering all round and enclosing a dense core of connective tissue with bundles of striated muscle which are continuations of the muscles of the proboscis. . . . Numerous groups of round laminated bodies in relation to the bundles of striated muscles in the core are noticed. They are a type of nervomuscular endorgan peculiar to the elephant. The three kinds of nerve endings found in the prehensile finger are found in the clitoris also. Such sensory structures are not found in the skin of the proboscis. Obviously the prehensile finger is the most sensitive and tactile part of the proboscis.

Here is evidence, if any more was needed, that an elephant's trunk is a sensuous organ. Neither is it a serpent nor a hand, as once imagined, but something much more erogenous. Despite moments of fantasy and blindness, scientific knowledge of the elephant has come a long way from the early misconceptions and embellishments of bestiaries like the *Physiologus*. Nevertheless, even the most clinical descriptions of an elephant's anatomy can accommodate human sentiments. At the end of his book, Professor Mariappa pays tribute to his mentor: "The special type of nerve endings reported in the prehensile finger of the proboscis and the clitoris are unique to the elephants. I wish to name these in honour of my preceptor and guide, Dr. A. Ananthanarayana Ayer, former Director, Institute of Anatomy, Madras, as Ayer's nerve endings."

III

my s o r e

Erstwhile Grandeur

$\mathscr{G}$uarding the entrance to the maharajah's palace in Mysore are the heads of two stuffed tuskers, one with its trunk raised and both with their ears flared, as if charging out of the wall to defend their territory. Of all the former princely states in India, Mysore enjoyed the greatest symbolic connection with the elephant. The Wodeyar dynasty, brought to the throne after the British defeated Tipu Sultan in 1799, adopted the elephant as an emblem of their authority. Mysore's royal crest features two elephant-headed lions in a rampant pose and the Wodeyars' palace is full of images and icons of tuskers. In the throne room hangs a portrait of one of the former rulers seated atop an elephant. This painting is exhibited in an oval frame formed out of the curve of two large tusks. The central gallery of the palace is decorated with an elaborate mural that depicts the annual Dussera procession, during which the maharajah was carried through the streets and around the palace grounds in a parade of elephants. Caparisoned in gold, with ornate howdahs on their backs and gilded parasols to shade their masters, these royal elephants epitomized the power and wealth of Mysore.

Today the tourism authorities re-create the Dussera procession using elephants hired from the forest department and private owners. The maharajah's golden howdah and other antique regalia remain on display, but the erstwhile ruler of Mysore no longer rides on an elephant. He lives in a private apartment at the back of the palace while the grand audience halls and galleries are open to the paying public. The Wodeyar dynasty, one of the richest in India before independence, has fallen on hard times, with escalating debts and unpaid taxes. After 1947, India's maharajahs lost most of their political authority, and in 1972 the privy purses, royal allowances paid by the government, were revoked. Much of the Wodeyars' land and other property have been confiscated or sold. Though the city of Mysore retains its name, the territories over which the maharajah once ruled are now part of the state of Karnataka. The royal stables at the palace contain no elephants, for the maharajah has no money to feed and maintain the animals that once elevated him to a position of power.

Yet the mythology of Mysore's elephants lingers on, with ivory relics and reminders of the days when Wodeyars still rode tuskers through the streets. The Karnataka state government emporia are full of elephants carved from sandalwood, rosewood, and marble—symbols of royalty reduced to souvenirs. In one of the curio shops in the center of the city, I found a six-foot-tall elephant sculpted from three huge blocks of teak. Hundreds of smaller images lined the shelves. Whether printed on postcards or carved into the handle of a paper knife, the elephant remains an emblem of the Wodeyars' reign.

For all its grandeur the Mysore palace feels empty and abandoned. The vast rooms are decorated in a mismatched style, with powder blue ceilings that look like Wedgwood china and heavy furniture that combines Victorian kitsch and Indian craftsmanship. Despite the priceless silver thrones and gilded howdah, there is a tawdry atmosphere to the capacious halls and a false sense of opulence. Looking at the mural of the Dussera procession, one can imagine the pomp and ceremony surrounding the festival, but there is also a sense of theatrical distortion. In those painted scenes on the gallery walls, the royal elephants are so heavily ornamented they hardly look like animals at all, obscured behind lavish costumes and makeup. Their physical features have been disguised completely and only their legs are visible beneath layers of fabric. Their faces, ears, and trunks are painted with floral patterns, and covering their foreheads are gilded faceplates with tassels and trim. Even their tusks are wrapped in gold bands. Seated atop these decorated beasts are the maharajah and his courtiers, including British officials who controlled most of the strings in this puppet state. The elephants themselves, however, seem to have disappeared behind the excessive trappings of a princely conceit.

Tango in the Jungle

Old milestones that mark the highway heading west from Mysore measure the distance to Kharapur, a small village that lies on the periphery of Nagarhole National Park. At one time this was the end of the road, beyond which there was nothing but jungle. On the outskirts of Kharapur, overlooking the Kabini River, stand two hunting lodges built by the maharajah of Mysore in 1929 for the viceroy's visit. Lord Irwin and his entourage traveled to Kharapur by Rolls Royce and Bentley to witness a khedah, in which a herd of wild elephants was captured. Also invited for this event were the maharajahs of Benares and Jodhpur, the maharani of Cooch Behar, and the prince of Persia. By all accounts it was a grand event and Krishnaraja Wodeyar IV spared no expense to entertain his guests at Kharapur.

Today the buildings where the princes and the viceroy once stayed have been converted into a wildlife resort operated by the Karnataka state government. Photographs decorate the walls, black-and-white images of hunting parties that visited Kharapur in the years before independence, including a Russian grand duke and an array of lesser colonial officials.

Across the river from Nagarhole National Park lies Bandipur Tiger Reserve and beyond that is Mudumalai National Park in Tamil Nadu state. Along with Wayanad Sanctuary in Kerala, these forests are part of the Nilgiri Biosphere Reserve, one of the largest surviving tracts of protected wildlife habitat in India. This region provides refuge for roughly 4,000 elephants as well as tigers and many other threatened species, from the diminutive mouse deer to the imposing gaur, a kind of wild cattle.

The Nilgiri Biosphere Reserve covers an area of 5,520 square kilometers, though sections are fragmented and elephant herds within these protected areas often come into conflict with farmers on the periphery. A telemetric survey conducted by the Bombay Natural History Society has shown that herds can roam over areas as large as 500 square kilometers, and the home range of these elephants often extends beyond the bounds of sanctuaries. The raiding of crops by wild elephants is a constant source of tension between the state forest departments and villagers living near the parks. When fields are threatened or destroyed, farmers often retaliate and a number of elephants have been wounded or killed.

Following my visit to Corbett Park, I traveled south to Mysore and joined my wife, Ameeta, at Kabini River Lodge. Though it was mid-January, temperatures were much warmer than in northern India and instead of being bundled up against the cold, we were able to drive around the park in shirtsleeves. The landscape and forests were very different from the foothills of the Himalayas. Driving from Mysore, we crossed over the last stretches of the Deccan plateau, with rolling farmland and dramatic outcroppings of boulder-strewn hills. Nagarhole Park lies at the edge of the Western Ghats, a range of mountains that separates the plateau from the Malabar Coast. The Kabini River has been dammed several miles below the lodge. In winter the reservoir is at its lowest level, exposing a grass-covered shoreline that looks like spacious lawns hedged in by thickets of bamboo and broad-leafed jungle.

Most of the forests near Kharapur are teak, planted by the maharajah over eighty years ago. These tall trees stand in ordered ranks and, despite the tangled undergrowth, give the forest a regimented appearance. Elephants often feed on the bark of teak trees, tearing off strips to supplement their diet of leaves and grass. "Debarking," as it is called by elephant experts, often damages and sometimes kills a tree. Scientists are not entirely sure why elephants feed on tree bark, possibly because of the fibers it contains or because of a concentration of minerals and other nutrients. A more staple food for elephants is bamboo, which grows in profusion throughout the park. The forest also contains a number of towering rosewoods and yellow teaks, as well as magnificent banyans that have laid out columns of rooting vines.

One of the unusual aspects of elephant behavior during the winter months in Kabini is the way in which the animals kick up the short grass along the riverbank before plucking it with their trunks. On our first afternoon at Kharapur, Ameeta and I took a jeep ride and came upon a herd of twelve elephants feeding in this way. Shuffling their front feet back and forth, they uprooted the grass with their toes, then used their trunks to gather it into a tidy mouthful. The elephants seemed to be moving in unison, and we felt as if we were watching a carefully choreographed dance, their rhythmic steps accompanied by the swinging of pendulous trunks.

Though the quantity of grass consumed in this way is obviously inadequate to satisfy an elephant's enormous appetite, it must be a seasonal delicacy that has nutritional value. Even the youngest calves were learning to kick at the ground with their feet by watching the adults and it looked as if they were eagerly trying to keep in step with the dance.

The myth of dancing elephants has been around for centuries, perpetuated by circus trainers, who are able to teach their animals to perform an awkward waltz or polka. Elephants do seem to have a natural sense of rhythm but the notion that they dance in the wild is, of course, absurd. More than any other writer, Rudyard Kipling fostered the myth in his *Jungle Book*. The inspiration for these tales probably came from his father, John Lockwood Kipling, who wrote a short book called *Beast and Man in India,* in which he recounts a folktale about dancing elephants. He writes, "Let us believe, then, until some dismal authority forbids us, that the elephant *beau monde* meets by the bright Indian moonlight in the ballrooms they clear in the depths of the forest, and dance mammoth quadrilles and reels to the sighing of the trees and their own trumpeting, shrill and sudden as the Highlander's hoch!"

Watching the elephants doing their synchronized two-step on the shores of the Kabini River, it is difficult to resist the romantic mythology of these dancing beasts. But as the herd moves together, swaying and shuffling about, they seem to follow their own subtle tempo rather than the drumbeat of human rhythms.

"I call it the elephant tango," says John Wakefield, resident director of the Kabini River Lodge. Known as "Papa" by friends and staff, Wakefield has spent most of his life in the forests of South Asia.

A retired Indian Army colonel, with a salt-and-pepper moustache, Papa is in his mid-eighties—as old as the teak trees that grow along the banks of the Kabini River. Dressed in a rumpled camouflage jacket, he seems most comfortable when driving through the jungle in his jeep. There is a distracted congeniality about Papa's manner, a raconteur whose stories have been told many times and are aged to perfection. His voice carries the lilt of an Anglo-Indian accent that one seldom hears anymore. As he speaks, he pushes his glasses up onto his forehead, where they settle for a few seconds, before sliding back down to the bridge of his nose.

Papa warns me, "Don't mind if I doze off in the middle of our conversation," but once he starts talking about elephants there is no danger of that. Learning that I have recently been to Corbett Park, he tells me that he once went fishing there, from the back of an elephant.

"There's a famous spot on the Ramganga called Champion's Pool where I had one of the mahouts take me into the middle of the river. The beauty of it is that an elephant doesn't disturb the fish when it wades through the water, and sitting on its back is perfect for casting a fly. I caught a five-pound mahseer that day."

Before I can ask a question, Papa launches into another story:

"They say an elephant has a good memory and I know that's true. When I was in my twenties, I was invited to go hunting at Gauri Chilla Block in the Terai. The chief engineer of the United Provinces . . ." It takes Papa a moment to remember the name. ". . . Stanton. He used to host an annual Christmas shoot, with dozens of guests. Each morning the mahouts would line up their elephants and we were given a chit of paper with our animal's number on it. As the hunt was about to begin, Stanton took off his pith helmet and tossed it up to the mahout. Immediately the elephant knocked Stanton to the ground with its trunk, then raised one foot, ready to crush him.

"Fortunately, the mahout dug his ankush into the elephant's ear to keep its head up. Everyone was paralyzed, not knowing how to react. Here was our host, about to be killed. All of us had rifles but nobody dared fire a shot. Finally one of the ladies in the group ran forward and opened her umbrella in the elephant's face. It was one of those bright-colored Japanese parasols, which frightened the elephant. He let go of Stanton and bolted, running for over a mile before the mahout finally got him to stop.

"Nobody knew why the elephant behaved like this. He had always been gentle and obedient but later we learned that his previous owner had been an alcoholic in Dehradun, who abused the poor animal. This man had a habit of throwing his topee up to the mahout before climbing aboard. Years later, when Stanton repeated this gesture, the elephant remembered and tried to take his revenge."

Like many former hunters, Papa is now an ardent conservationist. Wakefield and his colleagues have worked closely with the Karnataka Forest Department to develop and promote Nagarhole National Park. He points to a proportionately large number of tuskers as a sign of their success. With the pride of an indulgent uncle, Papa betrays a special affection for the animals of these forests.

"In Bandipur the elephants cause a lot of trouble but on this side of the river, they're well behaved. In the twenty-two years I've been here at Kabini, none of our guests have ever been harmed. Touch wood," he says, then raps his knuckles on his forehead.

Though the elephants in Nagarhole seem to live a charmed existence, dancing the tango on the banks of the Kabini, they often come in conflict with man. From his bedroom Papa brings out a framed photograph of a huge male elephant with crooked tusks that point in different directions like an errant moustache.

"A harmless old fellow," says Papa, as if he were showing me a family portrait. "Never hurt anyone. But he was found dead last year near the motor road. His tusks hadn't been cut off, so it wasn't poachers. The postmortem revealed that he died of an old bullet wound in his liver. Some farmer must have fired a shot at him while he was raiding the ragi fields."

Kabini Journal

1/11/2002
An hour before sunrise we set off in a jeep and drove through the forest to an inlet along the Kabini reservoir. Here we boarded a coracle, a circular boat made of bamboo and buffalo hide. Our

boatman paddled quietly past islands and sandbars, pausing from time to time and letting the coracle turn slowly in the water like a drifting leaf. The banks of the reservoir were studded with the stumps of teak trees, felled before the river was dammed. No animals were grazing along the shore, but there were plenty of waterbirds—egrets, ducks, cormorants, storks, and river terns. Except for an occasional riffle of feathers, nothing moved—a brief hiatus between night and day, as the birds, land, forest, water, and sky seem poised in anticipation of the sun. Unlike the winter mists that shrouded the Ramganga valley at Corbett Park, here the air was as clear as a polished lens, bringing everything into focus.

Paddling around an island, the boatman pointed out a crocodile lying in the mud a few feet away. Our coracle felt suddenly flimsy, nothing but a floating basket, though the mugger seemed to have no hostile intentions. When the sunrise cast its first rays on the shaggy thickets of bamboo, I noticed a bird hovering over the water. All at once, there was a commotion and a flurry of wings. Half a dozen river terns had taken to the air, crying out and harassing the interloper. As the bird jinked to avoid the terns, I realized it was an osprey. Circling sharply, then stalling for a second, it dove into the river and came up with a fish that dangled from its talons like a silver tassel.

1/12/2002

Driving through the forest in a jeep this afternoon we spotted a young tusker near the river, skulking about on his own like a petulant teenager. A short distance farther on we came to a clearing where the rest of the herd was grazing, kicking at the grass with their toes and swaying from side to side in unison. Most of the elephants ignored our presence, though we stopped the jeep only fifty meters away. An elderly cow, with a sunken forehead and furrowed skin that showed her age, eyed us cautiously as she fed. Despite legends and folktales of elephant kings, the leader of a herd is always a matriarch. Bulls usually break away from their families around the age of fifteen or twenty and live a solitary life, rejoining a herd only to mate. The young male that we saw first was probably getting ready to set off on his own. We watched the elephants for fifteen minutes, then started to move on, only to find that our jeep had a flat tire. The matriarch raised her head but showed no other signs of alarm as we got down from the vehicle, while the driver nervously changed the tire.

Soon afterward we passed another herd on the riverbank, almost identical to the first. Most were females, along with a few juvenile tuskers. On the opposite shore I could see herds of gaur that looked like tame cattle. Later, we came upon a pair of these animals at the roadside, one of them as big as a young elephant, with a massive neck and muscular shoulders. A large bull gaur can stand over six feet tall and weigh as much as 2,000 pounds. Their sharp horns curve up from a prominent boss on their forehead and the markings on their shins look like white kneesocks.

Driving on through the jungle, we found another herd of elephants beside a water hole, several kilometers away from the river. Again, the matriarch studied us warily as the others waded about in the scummy green water. One of them was three to four years old, a male calf whose tusks were just beginning to show. A large female, who could have been his mother, made a mock charge at our jeep to warn us off, after which she stood in the gathering twilight and faced us threateningly. Screwing up his courage, the calf sauntered forward, then lifted his trunk and gave a shrill cry before turning around, raising his tail insolently and farting in our direction.

1/13/2002

Setting out around four in the afternoon, we rode upriver in a motorized launch and saw over fifty wild elephants along the shore. There are so many here at Nagarhole it sometimes seems as if they outnumber the other animals. In one herd was a newborn calf, lying down between its mother's feet. The baby could have been only a few days old, and when it got up and stumbled about, several female elephants stood around in a close circle, touching it with their trunks. Cows in a herd often share responsibility for the protection and supervision of calves.

Farther upriver we came upon a lone male whose tusks were so long they crossed at the ends. He had obviously been rolling in the mud, for he was caked with dirt and dust, turning him a khaki color. He wasn't disturbed by our approach and the boat passed within twenty meters of him. The afternoon sun glinted off his tusks, which looked so heavy they seemed to weigh him down and must have made it difficult for the elephant to feed. The expression "long in the tooth" suited him perfectly, and the old bull's solitary presence made me think of a misanthropic recluse who has no time for socializing and could never be persuaded to dance.

Less than a kilometer ahead we passed more elephants, this time grazing beside mixed herds of gaur, cheetal, and wild boar. As the boat driver cut the engine and we glided toward the shore, the bison grew skittish but the elephants ignored us, except for one tusker standing by himself at the water's edge. He was feasting on weeds and glowered at us in disdain. His tusks were much shorter than those on the bull we'd just seen, pointing straight ahead and sharply tapered at the ends. He seemed to be in prime condition and in a fight he probably could have easily routed the old tusker. When we were thirty meters away, this elephant swung around belligerently. If he had wanted to, he could easily have capsized the boat. Instead he blundered up the embankment and uprooted a tussock of grass. Raising his trunk, he threw dirt at us, though most of it landed on himself.

1/14/2002

Driving along the unpaved forest road, we passed a ruined temple surrounded by bamboo thickets, where the guide told us leopards are sometimes seen. Last night one of the guests, who was riding with us in the jeep, saw a pair of tigers along the main highway from Mysore, just outside the entrance to the park. Near a water hole, we came upon a herd of elephants standing in profile, silhouetted against the green bamboo as if they were statues. Later we saw a Malabar giant squirrel leaping through the teak trees overhead like a furred acrobat. We also spotted a serpent eagle that had caught a snake. Crouched on the ground, the bird looked as if it were injured but took off with a shriek, the snake hanging from its claws like an untied shoelace.

Toward the end of our drive we came to a forest department compound in the middle of the jungle, where several tame elephants are stabled. Yesterday morning we were brought here for an elephant ride, but the animals had wandered off in the company of a wild herd and the mahouts had gone to find them. Now they were back in custody. The adults had been chained but one of the babies was loose and came up to the jeep, begging for something to eat. He was about three-and-a-half feet tall, with bristly hair on his forehead and small ears. Catching hold of my arm with his trunk, he was insistent, nuzzling me as I scratched his forehead. When he found that we had nothing to feed him, he moved to the front seat of the jeep and began chewing on the synthetic cover, making suckling sounds interspersed with groans of disappointment.

Khedah

In 1864 a young Englishman named George P. Sanderson set sail for India with ambitions of becoming a coffee planter. This career never got off the ground, however, as his arrival in Madras coincided with an infestation of borer insects that decimated the plantation where he had been offered a job. Following an extended period of unemployment and shikar, Sanderson finally secured a position with the Mysore irrigation department as an assistant channel superintendent. His interest in hunting and wildlife, however, dominated most of his energies. He later published a memoir titled *Thirteen Years Among the Wild Beasts of India,* which remains one of the most authoritative books on elephant behavior and captivity. In 1873 Sanderson persuaded the British authorities and the maharajah of Mysore to fund an expedition for capturing elephants. In his book he admits, "I knew nothing of elephant-catching at the time, nor had I any men at command who did; but I knew where there were plenty of elephants, and I was well acquainted with their habits."

At this time the traditional method of capturing elephants in southern India consisted of digging a deep hole in the ground and covering it over with bamboo, dirt, and grass, so that an elephant stepping on it fell into the trap. This pitfall technique was used by tribal hunters but had the disadvantage of being inefficient and injuring many of the animals. The method that Sanderson chose to use instead was khedah, originally developed in Bengal and other parts of north India, which involves the construction of a large stockade into which the animals are driven by beaters.

The area Sanderson selected for his first khedah was southeast of Mysore, in the Biligirirangan Hills. Setting up headquarters near the village of Morlay, he began to make arrangements that assumed the logistical dimensions of a military campaign, employing over three hundred men. A great deal of skepticism surrounded this operation, partly because of unsuccessful attempts in the past. Hyder Ali, a former ruler of Mysore, had led an expedition to catch elephants in the same forests in the 1770s but eventually gave up. On a stone marker in the jungles near Morlay, he carved an inscription that bitterly recorded his own failure and put a curse on anyone who tried to capture elephants in these forests.

Though the British authorities supported Sanderson's plans,

many of the colonial officials in Mysore dismissed the khedah as an expensive mistake. This did not dissuade Sanderson.

> I was determined to make the scheme succeed if possible, not only from my love of adventure, and the necessity for executing what I had suggested to Government and undertaken to carry out, but from the desire to prove to several officials who considered the scheme to be the vision of a lunatic, that their croakings were rather the utterances of Bedlamites. Pleasantries appeared in the Bangalore papers regarding the probable effect the kheddah operations would have on the price of salt, which it was represented was being laid in by me in large quantities for application to the caudal appendages of any elephants I happened to meet with!

In principle, the design and execution of a khedah was relatively simple but it involved careful planning and coordination. As Sanderson describes it, he first located a large herd of elephants in the forests bordering the Honhollay River, a tributary of the Cauvery. With his army of men, he surrounded the elephants and set up pickets who manned bamboo barricades and bonfires to keep the animals from leaving the area. At a suitable place, along a path frequently used by the elephants, Sanderson constructed a stockade, with two wooden palisades funneling into the enclosure. The entrance of the stockade was about twelve feet wide, with a heavy wooden gate suspended from above. This was raised by a pulley and rope so that the gate could be lowered after the herd was safely inside.

Once the stockade was built, the men who had encircled the herd of elephants drove them forward. With drums and flaming torches, the beaters closed in around the elephants while blocking all routes of escape. On the first attempt, however, the herd stampeded and broke through the ring of beaters. It must have been a chaotic scene, with frightened elephants charging past the line of men, all of whom were shouting and pounding on drums. To Sanderson's great disappointment the elephants crossed over the river and escaped. After waiting for several weeks, he got reports that the herd had returned and quickly marshaled his troops to encircle the elephants once again. This time Sanderson's men were better prepared and successfully drove the herd of fifty-three animals into the stockade.

After this the difficult process of roping and taming the wild elephants began. Mahouts entered the stockade on trained elephants and lassoed the animals one by one. They were then taken outside the enclosure and tethered to posts in the ground. The youngest elephants did not have to be tied, as they stayed with their mothers. One of the calves was an albino that Sanderson describes as being a "dirty cream" color. Water and food were provided for the elephants, but the ropes cut into their ankles as they struggled to get free. Gradually, they grew to accept their confinement and one elephant even allowed Sanderson to climb on its back after only six days.

His success in the khedah operation was trumpeted throughout the empire, and Sanderson was soon dispatched to Bengal, where he conducted several more khedahs. He later returned to Mysore, where he was given the title "Officer in Charge of the Government Elephant Catching Establishment." *Thirteen Years Among the Wild Beasts of India* contains descriptions of his adventures, as well as observations on elephant behavior. Having studied the Kannada language, he was able to converse with mahouts, and his book records plenty of elephant lore, including references to the three castes of elephants. Commenting on ceremonial events in Mysore, Sanderson writes: "Elephants are kept by natives of rank in India solely for the purposes of display and in this sphere the animal is more at home than in any other. The pompous pace of a procession suits his naturally sedate disposition and the attentions lavished upon him please his vanity."

Khedah operations became an annual event in Mysore, providing elephants for the maharajah's stables and generating revenue for the state. The *Mysore Gazetteer* of 1927 includes a chart that gives statistics between 1894 and 1918. The largest number of elephants captured in a single season was 170, during 1896–97. In a column labeled "casualties" the figure given is 52, proving that even though khedahs were a somewhat less brutal method than pitfalls, a large proportion of the elephants died. Adult bulls were usually shot inside the stockade because they were difficult to handle and train. Out of the 170 that were captured, only 79 were eventually sold and 39 "disposed of otherwise." On average, an untrained elephant was sold for Rs 1,379. Total revenues for the state were Rs 82,990, a considerable amount in 1897. Aside from procuring elephants for sale, the Mysore khedahs were also "undertaken to provide relief to the harassed raiyats, whose cultivation is destroyed by the elephant, or they may be ordered to provide entertainment to distinguished State guests."

On November 25, 1889, Maharajah Chamaraja Wodeyar X accompanied Queen Victoria's grandson, Prince Albert Victor, on a khedah organized by Sanderson near Kharapur. In his book *Modern Mysore,* M. Shama Rao provides an account of the event:

On arriving at the Khedda the party proceeded on foot outside the enclosure to a sort of jungle grand stand which overlooked at a distance of thirty yards the gateway through which the elephants were to be driven. . . . The Maharaja was entrusted with a knife for cutting the cord, an experienced hunter standing near to apprise when the correct moment arrived. . . .

Immediately the signal was given, the beat commenced and after much varied fortune, the herd breaking back more than once, the animals came and stood close to the tree on which stood the Prince's platform. His Royal Highness had

a good view of them here at the distance of but a few
yards. . . . At last in a compact herd, each individual elephant struggling not to be last, they crowded through the
gateway into the first enclosure urged on by several charges
of small shot which His Royal Highness plied them with.

Khedah operations at Kharapur continued to provide
"amusement" for a regular stream of visitors. The capturing of
wild elephants seems to have evoked a sense of drama and
adventure in the minds of colonial officials, perhaps because it
combined an elaborate administrative and logistical challenge
tinged with elements of tragedy—not unlike the empire itself.
Royal khedahs in Mysore ended around 1947, though the forest
department continued to conduct elephant-catching operations
intermittently after independence. The last of these was held in
1971 and attracted a large number of tourists, who purchased
tickets and were seated in a grandstand behind the stockade.
Once again the drama and adventure of catching elephants in the
wild was promoted as a spectator sport and the tourists witnessed a cruel finale to the khedah. As the herd of elephants was
being driven toward the gate, one of the older cows balked and
wouldn't let any of the others go forward. When the elephants
threatened to stampede, possibly endangering the tourists, the
matriarch was shot and killed, after which the rest of the herd
rushed into the stockade in a panic. Thus the final Mysore khedah
ended with an elephant's desperate act of resistance, answered by
a volley of gunfire.

Elephant Boy

The mythology of Mysore's elephants is not restricted to the exploits of maharajahs and colonial officials. There is also the story
of a young stable hand who went on to a career in Hollywood.
Here the legends and lore of the *Matangalila,* of Kipling's *Jungle*

Book, of G. P. Sanderson and the Wodeyars give way to the rags-to-riches tale of Sabu the elephant boy. Glossed over by his self-appointed narrators, details of Sabu's early life remain ambiguous. Born in Kharapur as the child of a mahout, he was raised among elephants. Sabu's mother died soon after his birth and one of the female elephants is said to have rocked the baby's cradle with her trunk. A few years later, following the death of Sabu's father, the boy was brought to the maharajah's palace in Mysore, where he lived as a ward of the state and learned to look after the elephants.

As with any myth, whether oral, written, or celluloid, it is important to recognize who is telling the tale. On one level, Sabu's biography represents the antithesis of imperial conquest and princely power—an orphan from the village of Kharapur who might have remained anonymous if it weren't for a fateful screen test in 1935. However, once Sabu was chosen to play the lead role in Robert Flaherty's movie, *Elephant Boy,* his life story was no longer his own. Instead, his identity merged with the fictional character in the film. As with so many other Hollywood stars, Sabu's origins were nothing more than material for a script and his story was always subject to revision.

Robert Flaherty specialized in making films with exotic settings and using "native" people as actors. He came to India after the success of his first movie, *Nanook of the North,* in which he narrated a seemingly realistic story of Inuit hunters in the Arctic. Financed by producer Alexander Korda, Flaherty set out to make a film version of Kipling's "Toomai of the Elephants," an episode from the *Jungle Book.* He was welcomed by British authorities in India and arrived in Mysore in the summer of 1935. Outfitted in a pith helmet and khaki safari suit, Flaherty undertook a film-shooting expedition that rivaled G. P. Sanderson's elephant-catching operations.

Though Toomai and his elephant Kala Nag (played by Irawatha, the largest tusker in Mysore) are the primary characters in this film, the story itself revolves around a khedah organized by a white hunter known as Carpenter Sahib, played by W. E. Holloway. The parallels with G. P. Sanderson are obvious, as the Englishman takes command of an ambitious elephant-catching operation, only to discover that the wild herds have disappeared. When Toomai's father is killed by a tiger, which Carpenter had wounded, the boy and Kala Nag run away into the forest to mourn. Asleep in the jungle, Toomai dreams of a huge herd of elephants that gather to dance in the moonlight. Waking up, he finds that Kala Nag has wandered off, and chasing after him, Toomai comes upon the elephants he saw in his dream. Soon afterward, Carpenter and his trackers find the boy, who leads them to the herd.

Flaherty's film contains all of the melodrama and cinematic sleights of hand that make for a commercial success. Produced during the Raj, it celebrates the colonial adventure and promotes racial and cultural stereotypes. Though Sabu plays the title character, he remains subservient to Carpenter Sahib and at one point even bows down in front of the white hunter and begs to be "beaten like a dog" for having disobeyed instructions. Many of the Indian characters, including Toomai's father, are played by British actors, their skin darkened with makeup. The dialogue is delivered in stilted English, even the commands that Sabu gives Kala Nag.

Most of the sequences from Toomai's dream were performed by tame elephants from the maharajah's stables. These animals were trained to raise their forelegs in unison like a clumsy chorus line. The most disturbing images in *Elephant Boy* are those that depict the khedah. Beaters run through the forest with flaming torches in hand, shouting and firing guns, while the wild elephants run about in panic, unable to escape. Two young tuskers blunder through bamboo thickets in obvious terror. As the herd rushes down the riverbank and into the swollen waters of the Kabini, one of the baby elephants is left behind, squealing with fear. Plunging into the muddy current, the calf looks as if it is drowning, while the others struggle to get across, unaware that a stockade has been built on the opposite shore. The captured elephants were released immediately after the khedah had been filmed, but they were clearly traumatized by the experience. Perhaps the worst cinematic deceit of all is the way in which several scenes shot during the khedah were spliced into Toomai's dream and set to music. Here the audience is led to believe the elephants are dancing, while they are actually packed together behind a wooden stockade and jostling back and forth in terror.

Once the location shooting was complete, Flaherty left Mysore and moved to London. Both Sabu and Irawatha sailed with him from Bombay to finish the film at Korda's studios in England. Irawatha completed his role as Kala Nag and was pensioned off to the London Zoo, while Sabu became a star. He soon moved on to Hollywood and acted in other movies, including *The Jungle Book, The Drum,* and *The Thief of Baghdad.*

Though Sabu spent most of his life away from India, he was able to make a triumphant return to Mysore in 1952. Maharajah Jayachamarajendra Wodeyar granted him a special audience and Sabu was welcomed by his many fans. India was now an independent country, the British had departed, the reign of the Wodeyars was all but over, and khedah operations had virtually ceased. Instead of riding into Mysore on an elephant, Sabu returned from America at the wheel of a Cadillac, which he brought with him by ship, a somewhat different emblem of wealth and power.

IV

remover of obstacles

The Island of Elephanta

*A*t the center of Bombay harbor, barely visible from the mainland, lies a thickly forested island with two rounded summits. Less than four kilometers in circumference, this protrusion of rock and foliage provides a verdant contrast to the concrete skyline of the city and the rusting hulks of cargo ships and tankers that wallow in the choppy waters of the bay.

Originally known as Gharapuri, Elephanta Island was renamed by Portuguese sailors who arrived off the western coast of India at the close of the fifteenth century. While pursuing opportunities for trade and plunder, they discovered an ancient cave temple carved into one of the two peaks on the island. Dating back to at least A.D. 700, this temple is dedicated to Shiva and contains sculptures of his many manifestations as well as a pantheon of attendant deities. The most famous carving at Elephanta is the giant Trimurthi, depicting the three aspects of Shiva—creator, preserver, and destroyer. Little is known about this temple's origins, for there are no textual references to the caves, and the only inscription that might have offered a clue was cut out of the rocks, carted off by the Portuguese, and probably lost at sea. Though the sailors from Lisbon were seemingly unimpressed by the spectacular Hindu idols, their imagination was captured instead by the stone image of an elephant that guarded the entrance to the temple.

Unfortunately, this monolith no longer stands on the island that bears its name. In 1865 it was removed to the mainland by British authorities and later placed on display near the zoological gardens in the heart of Bombay. Today the visitors who flock to Elephanta Island ride across the bay in motorized launches instead of sailing ships. As they climb the broad stone steps from the pier to the cave, tourists run a gauntlet of souvenir stalls and curio shops, offering innumerable images of elephants in all shapes and sizes. Carved from soapstone and marble, from rosewood and sandalwood, the various images of elephants line the approach to the cave like a ceremonial procession, saluting the original elephant who no longer inhabits the island.

I visited Elephanta toward the end of the monsoon, and the crossing from Apollo Bunder docks was rough, waves slopping over the bow of the launch. The water in the harbor reflected the metallic gray of the clouds and it began to rain as we approached the long cement pier. Mangrove trees grow in the tidal flats that ring the island, as if the forest has spilled into the sea. Though the drizzle of rain soon dissipated, the humidity remained high and even the rocks seemed to be perspiring. In addition to the main temple there are several smaller shrines cut out of the basalt ridge. A spacious promenade in front of these caves is overshadowed by a grove of tulip trees, known as bhendi, with paisley leaves and large drooping blossoms of yellow and red. The jungle, which covers most of the island, is maintained as a bird sanctuary. Though separated from the city, Elephanta is not completely isolated from the incursions of urban sprawl. Pollution from the bay washes up on its shores and seismic shocks from dynamiting on the mainland, where factories and high-rise buildings are being constructed, have damaged many of the temples.

Wandering beneath the canopy of leaves and through the pillared halls of the caves, I tried to imagine the armies of stone carvers who tunneled into these cliffs, extracting mythological images like sacred fossils mined from the solid depths of stone. It is safe to say that probably no elephant has ever set foot on Gharapuri, except through the collective imagination of those anonymous artisans. In fact, there is nothing that connects this tiny island to *Elephas maximus,* except the sculptors' vision and the misconceptions of European seafarers, compounded by several centuries of tourist tales.

With the statue removed, Elephanta is little more than a name, though the history and politics of nomenclature in India are infinitely complex. Only a few years back, Bombay itself was officially renamed Mumbai, in an attempt to rid the city of its colonial antecedents. It was the marauding Portuguese who first dubbed it a "good bay," a deepwater port that soon became one of South Asia's busiest harbors, and an important link with Arabia, Africa, and Europe.

Because of the changeable nature of names, I had to decipher several guidebooks and maps in order to find the elephant that once stood on Gharapuri island. The zoo in Mumbai, once known as Victoria Gardens, is adjacent to the Museum of Economic Products, originally the Albert Museum—not to be confused with the larger Prince of Wales Museum. Today the zoo is officially known as Veermata Jijabai Bhonsle Udhayan, and the museum bears the name of Dr. Bhau Daji Lad, a wealthy citizen of Bombay who contributed to its construction in 1848. The taxi driver who took me there, however, recognized none of these names, and after puzzling over them for some time, he finally realized that my destination was popularly known as Rani Bagh—the queen's garden.

Several wrong turns later, I found the elephant standing in a narrow patch of lawn, to one side of the museum. Sharing this space were a number of statues of English governors and military officers who had been removed from roundabouts or city squares and relegated to the grounds of the museum. These life-sized images, stained and weathered from years of exposure to the salt air and sun, are ranged behind the elephant like a deposed guard of honor.

The stone sculpture from Elephanta has none of the pomposity or human vanities of the colonial statues. On first impression, it is a very ordinary elephant, about five feet tall and carved out of the same dark basalt that gives the caves their shadowy, sepulchral aura. There is no decoration, no hint of captivity. The body is heavily formed, reflecting the contours of the huge boulder out of which it was shaped. The elephant has a ponderous and stoic dignity, standing in the garden beneath palm trees, herbaceous hedges, and a hibiscus bush with scarlet flowers. Over the years it has suffered damage, including the loss of one ear. (The Portuguese used the cave temples for target practice.)

Sprouting from a crack in the elephant's forehead is a sprig of green, a monsoon weed that took root in a plastered crevice and gives the animal a jaunty appearance. Though the carving on the body is minimal, with pillarlike legs, the face and trunk display an appreciation for the elephant's form, the way its trunk coils up beneath one tusk, as if feeding on plants in the garden. There is something enduring about this sculpture that holds our imagination far more than the bronze statuary and the Palladian architecture of the museum. It is older than any of the surrounding structures by over a millennium and has a primal, unfettered quality that makes one feel as if a living creature could easily emerge from the stone.

A freshly painted sign at the elephant's feet provides an unpunctuated explanation for its presence.

The originally stored about two hundred and fifty years to the right of the leading place for the celebrated cave on the island of Elephanta of Gharapuri and the former name was given to the island on account of the Elephant which was probably erected Originally at the same period as caves were constructed the date of which has been conjectually placed between 400 to 1000 AD was standing till the end of the 18th Century Early in the 19th century fell to places about 1864 largely owing to the efforts of Dr. (After words Sir) George Birdwood. The Fragments were removed to the Victoria Gardens (V.J.B. Udayan) where they were re-erected in the year 1914.

Reading this, one realizes how easily history becomes garbled and the futility of trying to establish infallible links with the past. Instead, the elephant itself seems to rise above the confusion of names, dates, and places, reminding us that man has always admired this beast no matter when or where it stood.

Ganapati

Every schoolchild in Mumbai is taught that Lokmanya Tilak, one of India's freedom fighters, organized the festival of Ganesha Chathurthi in 1893. It was conceived as a means of fostering unity among diverse religious communities, while avoiding restrictions against public assembly imposed by the British authorities. Over the years, this festival has become one of the most popular events in the state of Maharashtra and it is celebrated with enormous enthusiasm. Ganesha, or Ganapati, the elephant-headed deity and son of Shiva and Parvati, is worshipped for ten days in temporary shrines known as mandals that are set up in homes, on street corners, and in the narrow chawls, or neighborhood lanes, throughout the city. During this festival, which occurs near the end of the monsoon in early September, these Ganapati mandals become the focus of worship, where every day the elephant-headed god is feted and fed. Hymns are sung in praise of his benevolence and crowds line up to receive his blessing. On the final day of the festival, the clay statues of Ganesha are paraded through the streets, then immersed in the Arabian Sea.

Though most of the household idols are no more than a foot high, quite a few of the larger Ganapati mandals exhibit images over twenty and sometimes thirty feet tall. These are elaborately decorated, painted in bright colors, and ornamented with gilt and tinsel. Surrounding many of the idols are tableaus from Hindu scriptures and epics. What began as a simple ceremony that celebrated a common cause against the British Empire has become an important religious event as well as a public forum for popular demands and desires. Unlike many other Hindu festivals, it is loaded with secular symbols and contemporary references. For example, one of the Ganapati mandals featured the struggle against Pakistani terrorists, with a ferocious Ganesha wielding a sword and driving the infiltrators out of the country. Another tableau portrayed Ganesha as a judge in the trial of the actress-turned-politician Jayalalitha, who was in the dock at the time on corruption charges. Lokmanya Tilak probably never imagined that he was setting in motion such a powerful pageant, wherein the struggle for freedom has been supplanted by other agendas and other forms of rhetoric. All of this takes place within the context of a mythological narrative—the story of Ganesha—that dates back long before 1893.

Ganapati is a favorite deity of the middle classes and symbolizes wealth and prosperity. Many of the mandals promote commercial themes, and some hold raffles for cars and television sets, while one was even modeled on *Kaun Banega Crorepati?*—India's version of *Who Wants to Be a Millionaire?* Advertisements for everything from chewing tobacco to toothpaste decorate the banners and archways at the entrance to each mandal. There was even one Ganapati created entirely out of Pepsi bottles—a consumable deity. Though some commentators have decried the overt materialism of the festival, a synthesis of the sacred and profane seems almost natural. Part of the reason for some of the excess and competitive gimmicks is due to the expense of constructing these idols. A large Ganapati image can cost as much as 40,000 rupees and the additional decorations and tableau can run the bill beyond 50,000. To support the annual investments in each mandal, many of the neighborhood shrines impose entry fees to supplement offerings and subscriptions from within the community. Some are even financed by underworld figures from Mumbai's criminal mafia—including a mandal that featured a scale model of the Mysore Palace, reportedly paid for by Chota Rajan, a notorious gangster.

The Bollywood film industry has always played a significant part in Ganesha Chathurthi celebrations. Producers and directors sponsor Ganapati mandals and pray for Ganesha's blessings on the movies they make. Set designers and artists are employed to construct and decorate many of the more extravagant mandals with all of the glitz and glitter of Hindi cinema. The synthesis of celluloid myths and Hindu icons blends layers of fantasy and faith.

At Raj Kamal Studios, in central Mumbai, there was a Ganapati mandal just inside the main gate. This image of Ganesha was about eight feet tall, seated on a silver throne with bright turquoise upholstery. The mandal presented Ganapati in his most familiar form, with four arms and his trunk curving to the left. His potbelly was girdled with a cobra and he was heavily garlanded with strands of marigolds, roses, and jasmine. He was seated under a canopy of colored lights supported by ornate pillars decorated with the images of peacocks. Flower petals were strewn at Ganesha's feet, almost hiding his companion and vehicle, the rat. A mound of at least thirty coconuts had been placed in front of him, along with trays for worship containing flowers, vermilion powder, and sweets. The brightness of the lights and the dazzling colors made the mandal look like a film set, and when I began to take a photograph, the priest hurried out from behind the shrine and posed in front of the image, like an actor responding to a cue.

Earlier I had visited a more modest shrine erected by the Mumbai Municipal Employees Union. This idol was about two feet high, brightly decorated but without a throne or elaborate pavilion. Half a dozen union members were sitting on the floor inside the mandal but instead of worshipping the deity they faced a television set. A Hindi film was playing, the hero engaged in mortal combat with a gang of villains. Even the idol of Ganesha seemed to be directing his benevolent gaze at the action on the screen.

Filmi Interlude

A young boy is lost in the forest. His distraught father prays to Ganesha, pleading for his safe return. As the child, whose name is Raju, lies unconscious in the jungle, he is stalked by a leopard, but just as the predator is about to spring on him, an elephant appears. Trumpeting loudly, it charges the leopard, grabs it by the tail, and chases it away. The elephant then fills its trunk with water and sprays the boy's face to revive him. Three other elephants emerge from the jungle and accompanied by his four new friends, Raju finds his way home. By this time the father is so overcome with grief that he lies on his deathbed, calling out his son's name. When Raju appears the father clasps him to his chest and learning how the elephants rescued his son, he says that these must be incarnations of Ganesha, who have answered his prayers.

All of this takes place before the credits begin to roll for *Hathi Mera Sathi (My Companion the Elephant)*, a film made in 1971, starring Rajesh Khanna, Tanuja, and four young conscripts from the Great Oriental Circus. Projecting a larger-than-life presence, elephants have often appeared in Indian cinema. In early mythological and historical movies, they stole scenes from silent gods and black-and-white heroes. But only with the production of *Hathi Mera Sathi* did elephants get their first opportunity to play a major role in a romantic comedy.

As with most Hindi films, *Hathi Mera Sathi* combines romance, music, pathos, tragedy, action, and humor. The story continues as Raju grows up in the company of his four elephants, playing football with them and sharing in their teenage mischief. Eventually Raju falls in love with Tanu, the daughter of a wealthy businessman. It is the elephants who first bring the couple together when Tanu's car runs out of petrol near Raju's home. Coming to the rescue, they tow her car as Raju flirts and dances while singing the film's theme song, *"Chal, chal, chal mera sathi, O mera hathi."* (Carry on, carry on, carry on, my companion, O my elephant.) Soon after the two lovers get engaged, disaster strikes, as Raju loses all his property and money in a fraudulent court case. Faced with the choice of selling his elephants to clear his debts, he refuses and soon finds himself homeless and destitute. Tanu's father breaks off the engagement and what follows is a series of adventures in which Raju and his loyal elephants work their way out of poverty. Performing on street corners, they collect money and eventually start a circus.

The stunts in *Hathi Mera Sathi* are difficult to watch, particularly scenes in which the elephants walk on their hind legs, ride bicycles, or stand on their heads. Knowing the cruel methods of training in circuses and the pain these performances cause the elephants, it is difficult to accept the film's theme of companionship between man and animal. Whenever Raju is in danger, the elephants come to his rescue, fetching a doctor to treat his fever, chasing a thief who steals his money, and killing a cobra that enters his house. One of the noble elephants even sacrifices his life for his friend by stepping in front of the villain's gun. In the funeral procession, Raju drags a cart carrying his companion's corpse and sings a mournful ghazal about the affinity between man and beast, while human characters and elephants shed copious tears.

The deity Ganesha plays a prominent role in the film and in one scene the elephants are shown worshipping his image. They spray water on the idol, drape it with garlands, kneel down in front of Ganesha, perform a puja by ringing bells, and finally circumambulate the image. *Hathi Mera Sathi* contains many other references to the role of elephants in Indian culture, particularly as symbols of prosperity and power. But more than anything, it is their playful nature that sparks romance between the two lovers.

Patron Deity

Combining the trunk, tusks, and ears of an elephant with the limbs and torso of a man, Ganesha embodies a composite representation of divinity. Compared to the stern and forbidding carvings of Shiva that fill the caves at Elephanta Island, Ganesha's visage is cherubic and benign. He is a portly, playful god, almost comical in the juxtaposition of features. But in the minds of his devotees Ganesha is anything but fanciful, evoking potent myths and symbols. During the Ganesha Chathurthi festival, he is welcomed into the household as an honored guest whose presence brings joy and good fortune. Unlike more permanent statues of Ganesha, carved from stone or cast in metal, there is a transient quality about the clay images worshipped during Ganesha Chathurthi, an ephemeral brightness that survives for only ten days before they dissolve into the sea.

Several different accounts of Ganesha's birth can be found in Sanskrit texts like the *Puranas*. In some myths he is said to have been formed out of the oils and exfoliated skin of the goddess Parvati; in others he was immaculately conceived by Shiva through meditation. A more earthy version of the myth describes how Ganesha was conceived after Parvati and Shiva happened to observe two elephants mating in the forest. Forever curious in the art of love, the god and goddess transformed themselves into a tusker and cow so they could experience the same elephantine pleasures. Most accounts of his birth agree that Ganesha was born as a beautiful child, who acquired the head of an elephant under tragic circumstances. According to the most popular legend, Parvati asked her son to stand guard while she was taking a bath. When Shiva approached, the boy did not recognize his father and refused to let him pass. Angered by Ganesha's insolence, Shiva opened his third eye and burned off his son's head. When Parvati came rushing out to see what had happened, she was overcome with grief and her sorrow threatened to destroy the world. In a desperate attempt to console his wife, Shiva caught hold of the first living creature he could find, which happened to be an elephant, and he transplanted its head onto Ganesha's neck.

Another variation of this myth explains that when Ganesha was born, he was so perfectly formed that all of the gods came to admire him. The stars and planets also joined in these celebrations, but when Shani, or Saturn, looked at Ganesha, his evil eye decapitated the child. As before, Parvati became so distraught that the host of deities rushed out to find another head and the first animal they came upon was an elephant.

Ganesha is often linked to the sacred syllable ॐ (om), which suggests the shape of an elephant's head in the Devanagari script. The ears and trunk, as well as the protruding forehead, can be traced in the character, which is sometimes incorporated into modern, stylized representations of Ganapati. He is often shown holding a conch shell that is blown during worship to create the sound of om.

The iconography of Hindu deities is infinitely complex and Ganesha is no exception. Each element in a painting or sculpture of the god is loaded with meaning, from the objects he holds in his hands to the ornaments that adorn his image. Most representations of Ganesha depict him with four arms, though sometimes he can have as many as ten. In each of his hands he carries an emblem that suggests a story or an attribute. Two of these objects

are a rope and an ankush, or "elephant goad." The first is sometimes interpreted as a warning against greed, which is said to
snare a man like a noose. The goad is used to prod human beings
into following the correct path. More significantly, these two objects are also used to tame and train captive elephants.

Most images of Ganesha depict him with a broken tusk, and
he is often called Ekadanta, which means "one tooth." (Con-
versely, an elephant that has only one tusk is generally referred
to as a "Ganesha.") Again, there are several myths about how his
tusk was broken. In the most popular account, Ganesha transcribed the *Mahabharata* by breaking off his tusk and using it as
a pen while taking dictation from the sage Vyasa. A more elaborate story relates how Ganesha once gorged himself on a huge
quantity of sweets. On his way home that night he stumbled and
fell. When Ganesha landed on the ground, his stomach burst
open because it was so full. The moon, who witnessed this accident, began to laugh. In his irritation, Ganesha snapped off a
tusk and threw it at him. This story serves as an explanation for
the phases of the moon, which was sliced into segments by the
tusk. Having vented his anger, Ganesha then caught a cobra and
tied it around his waist to hold his stomach together, which is
why he is often shown with a snake girdling his midriff.

As a rule, Ganesha is not an angry god, though he sometimes
displays martial aspects. More often he is a sedentary deity, who
enjoys nothing more than feasting on the offerings his devotees
present him. According to one myth, Ganesha fought a battle
against an elephant-headed demon known as Gajamukhaasura,
who was harassing the gods. Gajamukhaasura was invincible because of a boon bestowed on him by a great sage, but Ganesha
was able to defeat the demon by turning him into a rat, which he
uses as his vahana, or sacred vehicle and companion. It seems
appropriate that Ganesha's vahana is a rat, which feeds on the excessive mounds of sweets prepared in his honor.

Ganapati's corpulence reflects stereotypes of merchants in

India, whose potbellies are often perceived as a symbol of their prosperity. During the Ganapati festival many of the sweet shops in Mumbai prepare a confection known as modak, a rice flour dumpling, which is Ganesha's favorite delicacy. In many shrines a fresh modak is placed in the idol's hand each morning. Though usually stuffed with grated coconut and spiced jaggery, modak fillings can also include dried fruit and nuts like raisins, apricots, cashews, and almonds. Many different varieties and sizes of modak are available, a few of them as large as coconuts. Modak can be flavored with mango essence, rose water, or saffron. Some sweetshops even offer a type of sugar-free modak for diabetic customers, but essentially the modak is a tribute to Ganesha's sweet tooth and the richer it is the better. Prashant Corner, a popular confectioner in Mumbai, advertises fifty-one varieties of modak, including one made with real gold that costs 1,200 rupees per kilo.

It is important to recognize that Ganesha, despite his trunk, his tusks, and his flapping ears, is not entirely an elephant. Instead he is a mythological deity through which human beings try to comprehend the power and mystery of creation. Idols of Ganesha present a divine paradox, superimposing the facial features of an elephant on top of a recognizably human body. (Scholars use the word *therianthropomorphic* to describe this kind of image.) The very fact that he defies a rational explanation suggests the incomprehensible nature of a god.

The myths of Ganesha are full of anomalies and contradictions, as are so many narratives of Hindu gods. Even the emblems that Ganapati holds in his hands often change from one sculpture or painting to the next. What remains constant, however, is that he is neither an elephant nor a man, but a combination of both. Our two separate species are melded together to create an anatomical riddle. Composite images of man and animal, like the centaur or the sphinx, are a common mythological device found in many cultures. An interesting aspect of Ganesha, however, is that he has an animal's head and a man's body, the reverse of composite creatures in most western myths.

Despite the obvious distortion there is an eccentric logic to the imagery of Ganesha, and seeing the idols that proliferate throughout Mumbai during his festival, it is possible to believe that our two species could actually inhabit the same skin. Part of the reason may be a basic sense of affinity that exists between man and elephant in India. Through the various images and narratives, we can appreciate this commonality. Ganesha's gluttony, for instance, reflects the huge quantities of food that an elephant requires, but it also suggests insatiable human appetites and desires.

Scholars continue to debate the meaning and mystery of Ganesha, who is considered a relative newcomer to the Hindu pantheon. The first clearly identifiable images of Ganesha only appeared in Hindu temples during the fifth century A.D. However, there is evidence from Sanskrit texts that he evolved out of an earlier deity, sometimes referred to as Ganapati or Vinayaka, who was part of the "household" of Shiva. Archeologist M. K. Dhavalikar has even suggested that Ganesha was originally worshipped as a malevolent and recalcitrant demigod, known by the epithet "Lord of Obstacles," whereas later he was transformed into a remover of obstacles.

Historian A. K. Narain argues that the popular image of Ganesha, as he is worshipped today, is actually a composite of early representations of sacred elephants and tutelary deities known as Hastimukhas and Dantins that merged together to create a wholly new deity.

> The most striking visual characteristic of Ganesa is no doubt his elephant's face. . . . The disagreement and diversity in the stories detailing the birth of Ganesa do provide a temptingly fertile ground for speculations, but the fact remains that none of the Puranic accounts tell us that he was

born with an elephant's head. And if accounts agree on one point, it is that the elephant's head was a grafting done to restore or produce a new life, and that the elephant's head was a borrowed item. Thus, prima facie, there is a very clear indication to posit a separate and independent identity of *the* sacred elephant, on the one hand, and that of a son of Siva and/or Parvati on the other, before the merger takes place.

Narain goes on to discuss evidence for early elephant cults, particularly in the Indus Valley culture (3000–1500 B.C.), where images of elephants appear on clay seals that seem to have served as animistic totems. Similar images appear centuries later in Buddhist mythology, when the "Excellent Elephant," or Gajatame, announces the arrival of the Buddha in Queen Maya's dream. This elephant goes on to represent the Bodhisattva in several Jataka tales, parables about the Buddha's former lives. At the same time, within Hindu iconography, the elephant is associated with the storm god Indra and often depicted as part of a fertility image of the goddess Lakshmi, showering her with water. In this way the evolution of Ganesha, from elephant and demigod to the complex, composite deity we recognize today, can be traced through many incarnations.

Both Narain and Dhavalikar refer extensively to artistic and numismatic evidence in support of their views, including an Indo-Greek coin from 50 B.C. that seems to bear the image of a man's body with an elephant's head. The Greek governors who remained in northwest India and Bactria after the conquests of Alexander appear to have had a fetish for elephants and imported these animals from the Mauryan Empire. Many Greek coins from this period were imprinted with the images of tuskers, a symbol of royalty and power. There are also gold and silver coins that bear the likeness of Alexander and his successors wearing elephant headdresses, somewhat like masks.

Whether or not the self-aggrandizing coinage of the Greeks had anything to do with Ganesha, his origins can be traced back in many different directions, from the earliest Indus Valley images to Hindu and Buddhist mythology. Whatever interpretations and connections are made, the historical antecedents of Ganesha are probably as varied and paradoxical as the mythological accounts of his birth.

Hornbill House

On a busy street near Salim Ali Chowk, not far from Mumbai harbor, is a modern office building that looks as if it might house a shipping company or government secretariat. This, however, is the headquarters of the Bombay Natural History Society (BNHS), the oldest private institution dedicated to the study and preservation of wildlife in India. Founded in 1883, the society held its first meeting at the Albert Museum, where the Gharapuri elephant is displayed. Over the years BNHS has published a number of books on the flora and fauna of India and maintains collections of butterflies, birds, beetles, and other specimens. The society has had a series of distinguished curators and honorary secretaries, including S. H. Prater, who published *The Book of Indian Animals* in 1948, which remains the primary field guide to wildlife in India.

The man whose name is most closely associated with the Bombay Natural History Society is Dr. Salim Ali, a renowned ornithologist who died in 1987 at the age of ninety-one. Those who have had any experience in the forests of India are familiar with his name and will likely have his books on their shelves. In a memoir, *The Fall of a Sparrow*, Salim Ali describes how he first entered the halls of the Bombay Natural History Society in 1908, carrying the remains of a yellow-throated sparrow that he was eager to identify. His uncle was one of the few Indian members of the society at the time. Overawed by the mounted shikar trophies and other "animal remnants" that filled the society's museum, he was welcomed by the honorary secretary, Walter Samuel Millard, who confirmed the identity of the dead bird in the boy's hand. This encounter and the encouragement that Salim Ali received from Millard and his staff inspired him to become an ornithologist. Fifty years later, he himself would become honorary secretary of the society and its president. Salim Ali's influence on the BNHS and conservation efforts in India is immeasurable, particularly as he steered away from the shooting and stuffing traditions of shikar—though he was admittedly a hunter himself—toward a more scientific and professional approach to wildlife research and preservation. In the process he was mentor to a generation of naturalists who are working in India today.

Hornbill House, the present headquarters of the BNHS, is named in honor of "William," the great pied hornbill who was the society's mascot from 1894 to 1920. Today there are no snarling trophies on the walls, though the upper floors of Hornbill House contain collection rooms full of mounted birds and insects. Many of the larger specimens are housed in a special Natural History wing of the Prince of Wales Museum next door, which makes for an interesting juxtaposition of India's artistic and biological heritage.

Ascending a flight of stairs to the administrative offices of the BNHS, I couldn't help but feel the contradiction of being in one of India's busiest and most crowded cities yet entering a space dedicated to the flora and fauna of its jungles. Before I had even introduced myself, one of the clerks offered me a box of sweets in celebration of Ganesha Chathurthi. He then delivered my card to the honorary secretary, and a few minutes later I was ushered in to meet J. C. Daniel, who carries on the society's tradition of congeniality. He served as curator of the society before taking up his present post and has been chairman of the Asian Elephant Specialist Group, founded in 1976. This organization led to the first comprehensive elephant census in India and the creation of Project Elephant, a central government agency that provides direction and funding for conservation and research throughout the country. Daniel has also edited two books on the elephant, drawing together many articles that have appeared in the BNHS journal. Having just returned from a conference in Kolkata (formerly Calcutta), titled "The Call of the Elephant," Daniel told me that there had been over two dozen papers presented, including reports on the status and distribution of elephants in different parts of India. Clearly, there is a community of concerned and committed naturalists in India who have taken up the cause of the elephant.

As Daniel explained, the primary issue in saving the elephant is the protection of its natural habitat as well as forest corridors that permit the migration of herds from one area to another. By protecting the forests that sustain the current population, we also protect the entire ecosystem, so that the survival of the elephant means the survival of other endangered species, including insects, birds, or rare varieties of trees and plants. As the Bombay Natural History Society has demonstrated through its research and publications, there is no species that lives in isolation. Instead, policies for the protection of wildlife must take into account not just a high-profile creature like the tiger or elephant but also something as easy to ignore as a yellow-throated sparrow.

Even more important, for elephant advocates, is an appreciation of man's place in the environment. It is easy to say that human beings have destroyed and desecrated much of nature's abundance and beauty, yet if we are to salvage what little remains, the needs of human beings must be taken into account. In a country like India, with its vast population, there is a recognition that human beings cannot simply be excluded from a forest. In many cases there are indigenous communities who depend on forest resources for their livelihood. In other instances the spread of cultivation has brought farmers into greater conflict with animals like the elephant. For this reason, as Daniel points out, "Conserving the elephant means conserving the human environment." Though national parks and sanctuaries provide areas of forest in which wildlife can survive, the people who live on the periphery of these protected zones can never be totally separated from the animals within. The challenge is to balance the needs of human beings with those of other species.

On the wall of J. C. Daniel's office is a framed photograph of Salim Ali with a trim white beard and beaklike nose. His profile is reminiscent of the birds he studied, not quite a hornbill but certainly not a sparrow. Though an ornithologist by training, Salim Ali was committed to the broader mission of the BNHS, which seeks to conserve the environment as a whole. In his memoir he relates a story that illustrates this point. On a bird-watching expedition in Kerala, he was accompanied by an eccentric Austrian named Baron Omar Rolf Ehrenfels. As Salim Ali describes it, "We were stalking single file along a narrow animal trail through dense tall grassland about five feet high—the right kind of habitat for the Broad-tailed Grass Warbler *(Schoenicola platyura)*, on which my thoughts were bent." Suddenly, Ali and Ehrenfels came face to face with a tusker and both men immediately turned tail and ran for their lives. The baron quickly outdistanced Salim Ali, who finally caught up with him, only to be asked, "What was it?" Though they had gone in search of a rare bird that weighs only a couple of ounces what they found in the forest was a four-ton mammal, who shared the warbler's habitat.

In Procession to the Sea

During the ten days of the Ganapati festival, Mumbai is animated by the ebullient spirit of Ganesha. A mood of celebration permeates every corner of the city, from five-star hotels to humble chawls. Even the smallest Ganapati mandal is lit up all night with strings of colored bulbs, while hymns to Ganesha are broadcast over loudspeakers. On the fourth and seventh days there are minor processions, when some of the idols are taken to the sea, but the major immersion, or Ganapati Visarjan, occurs on the tenth and final day. Around noon the parades begin, as images are ceremoniously removed from their shrines and carried through the streets. Most of the traffic in the city comes to a halt and the roads are clogged with celebrants. Government offices, schools, colleges, and most businesses are closed, except for sweetshops and roadside hawkers selling pinwheels and bright cardboard noisemakers.

The sound of drumming reverberates throughout the city as percussionists from wedding bands accompany the parade of deities. Bursting firecrackers are amplified by the tall buildings on either side, adding anarchic rhythms to the already deafening thunder of drums. Shouts of *"Ganapati Bappa Morya!"* ("Praise to Lord Ganapati!") from the festive crowds are barely audible above the pulsing roar. Yet despite the earsplitting noise and violent explosions, the mood on the streets is cheerful and exuberant.

Each of the idols is accompanied by celebrants from a particular mandal. The smaller Ganapati images are transported on handcarts or in palanquins. Larger idols are taken to the sea in trucks or on giant floats that are pulled by dozens of men dragging long ropes. Knots of dancers lead the way for each of the idols, their faces and clothes dusted with red powder, thrown in celebration. Most of the dancers are young men who hurl themselves into the rhythm of the drumming, arms waving in all directions. Women also join in the dancing, for the festival is a family affair, with children riding on the carts or carried on the shoulders of their parents.

Though Ganesha Chathurthi is predominantly a Hindu festival, efforts are made to include all of the other religious communities in Mumbai. During the festival, newspapers run articles about Christians and Parsi who celebrate Ganesha Chathurthi and even place images in their homes. Among most Muslims there also seems to be a level of acceptance. One of the most famous Ganapati

images, known as Lalbaughcha Raja (King of Lalbaugh), is especially famous for the granting of wishes, and his procession follows one of the longest routes, taking twenty-four hours to reach
the sea. The idol travels through several neighborhoods that are
predominantly Muslim and in the evening, when the call to prayer
is broadcast from mosques, Lalbaughcha Raja's procession halts to
allow Muslims to pray. After the namaz is completed, Muslims and
Hindus greet each other before the parade continues. In a city that
has seen more than its share of violence between these two religious communities, this festival provokes rare gestures of unity.

Ganesha may be regarded as the remover of obstacles but his
processions bring the city to a standstill. Approaching the major
beaches, long queues of idols are barely able to inch forward
through the mass of people who congregate in the streets. Police
attempt to direct traffic and barriers are set up, along with elevated platforms from which authorities monitor the procession.
Boy Scouts and Girl Guides are pressed into service, lining sections of the route and holding hands to form human chains,
which are quickly broken in the crush. With their berets and
knotted scarves, the scouts seem to be playing a game with the
heaving crowds that jostle them from side to side.

At several points along the route of the procession, viewing
stands for local dignitaries are erected, with bright-colored canopies and folding chairs arranged in rows. Painted banners are
strung from balconies, offering felicitations from political parties
of all persuasions. Posters of grinning parliamentarians are everywhere, offering pious messages in praise of Ganesha. Neighborhood committees set up stalls where the dancers quench their
thirst with free glasses of juice and water.

As I made my way through the streets, I recognized several
Ganapati idols I had seen earlier when visiting neighborhood
shrines, including a dancing figure that looked as if it was going
to topple off its float. Virtually every image was different. One
Ganesha was ensconced on a throne of cobras, a five-headed Na-

garaja opening its many hoods above his head like a venomous umbrella. Others were seated under tasseled parasols, as if to protect them from a monsoon shower. Each of the larger floats seemed to be competing with the rest for inventive imagery. Humor is employed in many of the tableaus, and one huge Ganapati was lounging on a palanquin shouldered by a dozen rats, all of whom had protruding bellies and were wearing white Gandhi caps. The skin colors of the Ganapatis ranged from bright vermilion to pink flesh tones or livid blue. There was even a green Ganapati, who presented an "eco-friendly" image, his figure formed out of leaves. This particular float promoted an environmental message with slogans against deforestation and pollution.

A larger-than-life statue of Lokmanya Tilak stands in a park near the entrance to Girgaum Chowpatty beach, surveying the festivities he initiated over a hundred years ago. A broad crescent of sand extends from the foot of Malabar Hill to the seawall protecting Marine Drive, about a kilometer in length. From one end of the beach to the other spread a throng of people, so that the sand itself disappeared and the crowds extended into the surf. Food sellers and hawkers did frenetic business, hemmed in by hundreds of thousands of people who had gathered to watch the idols of Ganesha complete the final stage of their ceremonial journey.

For the larger statues it is a slow, ponderous process, crowds parting as the images are hauled across the sand. The heavy floats inch their way forward as the tide comes in to meet them. Often it takes several hours to complete an immersion, as each Ganesha sits in the shallow waters, like an enormous bather, slowly dissolving in the waves. The smaller images are lifted off the carts and final prayers are offered. One of the worshippers then carries the deity into the sea, wading out as far as he can go before releasing Ganesha into the water. Within the crowd there is a mood of jubilation and regret, for the ten days of celebration are almost over and another year will pass before Ganapati returns.

Overhead a navy helicopter hovers to keep an eye on the splashing crowds. Lifeguards riding Jet Skis and motorboats circle back and forth, waiting to rescue anyone who might wade out too far. As the sun goes down over the Arabian Sea and the sky darkens, tube lights flicker on and much of the beach is lit up with a fluorescent glow. During Ganesha Chathurthi it is considered bad luck to see the moon and many people walk with their heads bowed to avoid any possibility that this might happen. Fortunately, the monsoon skies were overcast and the only light came from the electric bulbs on the beach and the bright galaxies of Mumbai's skyline.

Later in the evening, as I retraced the route of the procession, crowds remained packed together, and the idols seemed to be completely stalled. Firecrackers continued to explode and the drumming made the air vibrate between the high-rise buildings. At a few points it was so crowded I had to circle through back streets and alleyways, where the spectators thinned out and the lights were less dazzling, but each time I returned to the main procession, there was animated celebration, which continued long past midnight.

The next morning, around seven o'clock, I made my way back to Girgaum Chowpatty. By this time the streets were empty and the banners and strings of lights sagged overhead. Sidewalks were littered with scraps of colored paper and tinsel, discarded noisemakers, and crushed pinwheels. A few cars and taxis moved along the streets, driving over the charred remains of fireworks that looked like fallen leaves after a hurricane. All of the shrines lay empty, with only the painted backdrops and canopies in place, their idols long since removed.

Arriving at the beach, I could see a last group of worshippers completing the final immersion. By this time most people had already gone home. An obese, saffron-colored Ganapati was being escorted out into the water by a couple dozen men. They were fifty feet from shore, and the bright color of the idol contrasted with the leaden clouds and the murky brown of the sea. Farther out were the remains of another image, a blue-skinned Ganesha with ten arms, tilted to one side and gradually breaking up with the outgoing tide.

Police were guarding the entrance to the beach and when I tried to go down to the shore, they waved me back officiously. Already there were cleanup crews working on the sand, collecting trash that the crowds had left and the remains of the idols, which had washed up on the shore in waves of gaudy flotsam. Garbage trucks were being loaded with piles of palm fronds and garlands of marigolds that littered the beach. Circling around one of the bamboo barricades, erected to control the crowds, I was able to slip through the cordon of police.

Near the statue of Lokmanya Tilak stood a group of college students wearing matching T-shirts with the panda logo of the World Wildlife Fund. When I spoke to them, they told me that several environmental organizations had banded together to help pick up debris and reduce the ecological damage of the festival. These student volunteers seemed undaunted by the task in front of them. As I watched them fan out across the beach, they seemed buoyed on by their idealistic purpose as well as the lingering excitement of the festival. Chowpatty looked like an environmental disaster area, the detritus of floats scattered across the sand.

By this time the saffron Ganesha had reached a point where the water was above his waist. Giving him a final push, the worshippers folded their hands in farewell and began wading back to shore. He was listing slightly but remained upright. Before I could watch him dissolve completely, however, a squad of policemen came by and escorted me off the beach. Only the cleanup crews were allowed to remain. Glancing back over my shoulder, I saw that Ganesha had begun to crumble in the waves, and now only his head was visible as he slowly returned to the sea.

V

questions of captivity

Feeding Time

𝒯heppakkadu Elephant Camp in Tamil Nadu's Mudumalai
Wildlife Sanctuary is home to twenty-seven elephants that are
cared for and fed by the forest department. Some of these are
working animals; others have been injured or orphaned in the
wild. There are several "pensioners" as well, who have served the
forest department for years and reached the age of fifty-five
when they are no longer required to work. The feeding of these
elephants involves carefully prescribed nutritional formulas, determined by an animal's age, size, and daily workload. A special
dining area has been constructed at the camp, with a cement
shed where food is weighed and prepared under the supervision
of a veterinarian. Displayed along one wall is a chart that lists
the names of all the elephants in the camp, their ages, dates of
birth or capture, and whether they are still working or not. The
food provided for each elephant is also listed—ragi (millet),
horse gram, salt, and jaggery (raw sugar), which the animals receive in the evening to supplement their daily diet of roughly
200 kilos of leaves and grass. Surrounding this shed is a fence
made of logs where the elephants line up every evening to receive their rations.

I had made no arrangements for my own accommodation or
meals at Mudumalai, but on reaching the sanctuary I was di-

rected to Bamboo Banks Guest House in the nearby village of Masinagudi. Though I arrived without reservations, a room was available and lunch was served within half an hour, a meal that made me wish I had the appetite of an elephant. Only five of us were eating—Zerene and Siasp Kothavala, the owners of the guest house, an English couple, and myself. More than a dozen different dishes filled the buffet, including mutton biriyani, crab curry, goat's brain curry, several varieties of vegetables and lentils, all served with appam, a kind of rice flour crepe. In addition to this there was a western menu of pork chops and roast potatoes. I ate and ate, then sat there stunned as the cook brought out an enormous tiramisu pudding for dessert.

Between repeated helpings of food, I learned that Siasp Kothavala is a retired tea planter. The Nilgiris, or "blue mountains," which rise up from the edge of Mudumalai Sanctuary, are famous for coffee and tea estates, many of which are owned by the Bombay and Burma Trading Corporation (BBTC), Kothavala's former employers.

A blustery Parsi with a broad, well-lined face, Siasp Kothavala speaks in a crusty Doon School accent. Even if he hadn't told me, I would have guessed he was a tea planter, for he fits the incorrigible stereotype. As soon as he learned that I was writing a book about elephants, he said, "I hope you're not one of those boo-hoo sob sisters."

Uncertain what this meant, I shook my head, though his opinions became clearer as he launched into a tirade against animal rights activists. A hunter in earlier days, Kothavala told me that he had "studied most wildlife down the barrel of a gun." He went on to complain that many westerners who come to India look at animals through a sentimental lens. "They've turned nature into a religion," he said.

Much of his scorn was directed at those who protest the treatment of elephants at Theppakkadu. Kothavala explained that these "sob sisters" don't realize that an elephant, no matter how well it is trained, needs to be handled with a certain amount of force, and sometimes chains are necessary. The Bombay and Burma Trading Corporation used to maintain hundreds of elephants for its logging operations, but these animals are no longer required. One of the elephants at Theppakkadu, a female named Cauvery, used to belong to the BBTC.

"I'm a great fan of the elephant camp," said Kothavala, describing how their veterinarians had recently struggled unsuccessfully to save the life of a wild tusker, shot in the jaw by a farmer protecting his fields.

Toward the end of our meal, Kothavala threw up his hands in a gesture of exasperation. "The problem with elephants is obvious," he said. "There are too many people and not enough land. Forests are being destroyed and the animals don't have the habitat they need."

Though recent encroachments on forest land are a problem, much of the original territory of the elephant was taken over by coffee and tea estates in the nineteenth and early twentieth centuries. These plantations extend throughout the Nilgiris, from the borders of Mudumalai Sanctuary to altitudes of four and five thousand feet above sea level. At one time all of this area was jungle, which has been usurped by spreading tea gardens that look like a continuous maze of manicured hedgerows. The bright green bushes, plucked to an even height, conform to the contours of the ridges. Tall shade trees with their branches pruned are scattered throughout the estates, lone survivors of the forests that once covered these mountains.

It is no coincidence that tea and elephants are found in close proximity, not only in the Nilgiris but also in Assam and Sri Lanka. A moist submontane environment, ideal for growing tea, is also perfect for the various species of plants and trees that these giant herbivores favor. *Elephas maximus* has one of the most eclectic diets—scientists have counted over a hundred species of plants and grasses on which they feed—allowing them to adapt to changes in the environment. More than anything, however, they require a substantial range of territory over which to browse and forage. Elephants will eat most grass and foliage but tea leaves and coffee beans are not something they savor.

Picturesque though the tea estates may be, there is no more dramatic evidence of man's influence on the environment than these carefully cultivated gardens. The elephant's natural habitat, an array of different species of bamboo and other grasses, as well as staple forage trees like *Kydia calycina* or *Grewia tiliaefolia* and wild delicacies including wood apple and tamarind, has been almost entirely supplanted by uniform acres and acres of plantations. This form of human encroachment may not exhibit the ugliness and devastation of other assaults on the environment, such as mining or dams, but the consequences for animals like the elephant are equally severe. Fortunately, most of the remaining forests in this region are now under the protection of the Nilgiri Biosphere Reserve.

Ancient precedents exist for conservation in Tamil Nadu. Historically, the kings of this region used to maintain sacred forests or groves in which one tree was designated as the "guardian tree." It was the duty of the king to protect this tree and the legitimacy of his reign depended upon its survival. Those who sought to threaten a king's rule would often try to chop down his guardian tree. One of the surest ways of protecting the tree, as well as the throne, was to tether a royal tusker to its trunk. As living symbols of the king, both the tree and the elephant were linked together, and in classical Tamil poetry this serves as a metaphor of power. Another method of contesting the authority of a king was for a challenger to tie his own elephant to the guardian tree, an act of defiance that could not be ignored. In this way, recognition of the essential relationship between elephants and the forest serves both as an ecological and a literary device.

Late in the afternoon, following a long nap to sleep off the effects of our enormous lunch, I headed back to the Elephant Camp to watch the animals being fed. Situated along the banks of the Moyar River, the camp lies directly opposite the headquarters of Mudumalai Sanctuary, where some of the tame elephants give joyrides for tourists.

Wandering down to the edge of the Moyar, a languorous stream that twists its way through the foothills of the Nilgiris, I watched two elephants being bathed. One was an elderly female who seemed reluctant to get wet, complaining and moaning as her mahout pushed and coaxed her into the river. When the cow finally lay down, she let out a groan of resignation. The other elephant, a young male, was much more willing to plunge in and obviously enjoyed having his back scrubbed with a coarse brush. Rolling from side to side, he blew bubbles with his trunk.

I was absorbed in watching these two animals when suddenly there was a loud explosion behind me, like a rifle shot. Turning around with a start, I saw a massive tusker standing about five feet away. He had just broken a stem of bamboo, which had made the noise, and he was feeding on the leaves. The elephant, obviously one of the tame residents of the camp, had come up behind me so silently I hadn't noticed. We were close enough for him to have reached out and tapped me on the shoulder with his trunk. Scrambling up the riverbank, I got out of his way. The old bull, whose tusks converged at the end, broke several more culms of bamboo, each with a resounding crack. For the next half hour I followed him slowly downstream, before he waded across the Moyar and headed toward the Elephant Camp.

It was now five o'clock and the mahouts had already measured out portions of food. The mixture of ragi paste and horse gram was molded into rectangular blocks and arranged in neat rows on cement tables in the feeding shed, somewhat like a cafeteria. Soon afterward the elephants began to arrive, several on their own and others with handlers on their backs. Most of the mahouts at Theppakkadu are members of the Irula tribe, for whom the Moyar valley has always been home. As an indigenous forest-dwelling community the Irulas have captured and trained elephants for centuries.

Before the meal was served, three of the youngest elephants performed a puja at the temple nearby. A small shrine with whitewashed walls and an orange-and-yellow roof, it is built into the roots of a ficus tree. Appropriately, the temple is dedicated to Ganesha. The three supplicants and their mahouts approached the shrine and each elephant kneeled in front of the idol. They were then given brass bells to hold in their trunks and circumambulated the tree while ringing the bells. In the lead was a ten-year-old female named Rajeshwari, a beautiful young cow with a ruddy complexion. She knew every step of the ritual by heart. Returning to the front of the temple, Rajeshwari bowed once more, as her mahout applied a tilak of sandalwood paste and vermilion on her forehead. The other two elephants, a pair of four-year-old males named Palam and Sathya, followed obediently behind. A group of about thirty spectators, tourists and forest department employees, watched the puja and afterward offered prayers of their own.

By the time I got back up to the feeding shed, at least a dozen elephants were waiting patiently behind the log fence. Among this group I recognized the tusker who had startled me near the river. His name was Indhar and according to the chart on the wall he was fifty years old. Next to him stood Rajeshwari. With the red tilak on her forehead she looked like a demure bride, her tiny ears pressed flat against her head and her trunk wrapped coyly around one of the fence posts. At first Indhar seemed to ignore her but then, with the lecherous insouciance of an aging lothario, he turned and slid his trunk between her legs.

Male elephants often do this to check if a female is in estrus, but Rajeshwari was still too young for mating and ignored Indhar's advances. He, on the other hand, became aroused by her presence and soon developed an elephantine erection, which looked like a second trunk protruding from his wrinkled loins.

Once the mahouts began feeding their elephants, however, all thoughts of sex were quickly forgotten. Rajeshwari received twelve kilos of ragi flour and two kilos of horse gram, as well as a handful of salt and jaggery, while Indhar, who was at least twice her size, got twenty kilos of ragi and six of horse gram. All of this was kneaded together into huge balls of dough that the mahouts carried out to the elephants and stuffed into their mouths. The feeding was supervised by a forest department official and several uniformed guards, though the mahouts did all the work. The gray and glutinous mixture looked anything but appetizing, yet the animals eagerly lifted their trunks and slurped up the food with relish.

Moorthy

One of the first things that happens to an elephant after being taken from the wild is that it is given a name, which not only identifies the animal but also creates a sense of intimacy and association in the minds of its captors. Names allow us to tell an animal's story and sometimes to call the animal our own. The list of elephants painted on the wall of the feeding station at Theppakkadu includes many names of gods and goddesses from Hindu mythology. There is also one resident of the camp identified as Moorthy, a makhna or tuskless male, named in honor of Dr. V. Krishnamurthy, retired director at Theppakkadu and one of the leading veterinary experts on elephants.

For some, however, the elephant named Moorthy bears another name. To them he will always be known as Loki, a messenger god from the frigid latitudes of Norse mythology. This name was given to him by American animal rights activists with the India Project for Animals and Nature (IPAN), which operates an animal shelter near Mudumalai. They have been quoted as saying that Loki is "the messenger to the world about the plight of the world's elephants." Unlike Kalidasa's cloud messenger, who communicated lyrical missives of romantic love, Loki has aroused a different kind of passion. Further parallels can be found in the Norse myth, for Loki is described as a giant among the gods and known to be tempestuous and destructive. Ultimately he is chained down and tortured by a serpent whose fangs drip with venom. As he writhes in pain and struggles to break free, Loki is believed to cause earthquakes.

The story of this makhna highlights many of the issues and problems associated with capturing wild elephants. At the same time it brings into focus the tensions and miscommunications between international conservationists and state agencies in India. The Tamil Nadu Forest Department captured the makhna on July 21, 1998, after he was declared a rogue. According to state officials, this makhna had killed a dozen people and caused substantial damage to crops in Gudalur district. He was tranquilized and brought to Theppakkadu with the help of trained elephants. In the process of being captured, the makhna was injured, primarily by ropes and chains that were used to restrain him, though he also suffered from earlier gunshot wounds inflicted by farmers.

Soon after the makhna arrived at Theppakkadu, IPAN became involved in the treatment and care of this elephant, whom they christened Loki. For the first few months, the activists helped provide medical assistance and food in cooperation with the forest department and the mahouts. Their relationship soon soured, however, because of conflicting philosophies of animal rights and wildlife management. IPAN advocated a gentler and less restrictive approach to the makhna's care, while the forest department still considered him dangerous and treated him as such. This confrontation came to a head when IPAN was accused

of interfering and their staff were banned from entering Theppakkadu camp or providing care to the elephant. The activists responded by surreptitiously taping a training session where the
makhna was confined in a narrow enclosure and beaten with
sticks when he resisted.

IPAN released the tape to the press and enlisted the help of
U.S. Congressman Sam Farr of California. The Indian ambassador in Washington was presented with an appeal detailing the
makhna's mistreatment and demanding that Loki be handed over
to IPAN for medical care and rehabilitation. As this highly emotional cry for animal justice was reported in newspapers, on
television, and over the Internet, Maneka Gandhi, a member of
Parliament, initiated an investigation. Gandhi is known as an out-
spoken advocate of animal rights and heads an organization
called People for Animals. Through her influence the Indian government has banned the use of trained bears, lions, and tigers in
circuses, and she has fought against the use of whips by jockeys
on racecourses. Under the circumstances, few people in India
would seem to have had greater sympathy for the makhna's
plight. Maneka Gandhi took a personal interest in the investigation but soon came to the conclusion that IPAN had exaggerated
and sensationalized the situation.

Another person drawn into the fray was the elephant expert
Raman Sukumar. He visited Theppakkadu twice after the accusations surfaced and in April 1999 responded to IPAN's allegations, reporting that most of the makhna's injuries had healed.
"He has been treated well, fed adequately and shows no signs of
trauma," wrote Sukumar. "I can assure you that the local tribal
people who act as elephant mahouts have lived for generations
amidst wild and captive elephants.... I see no reason to question their love or compassion for elephants."

Much of this controversy has now died down and the
makhna remains at Theppakkadu, part of the captive herd. When
I visited in January 2002, the makhna was one of the elephants

brought to the feeding station and, to my untrained eye, he looked healthy and docile. Some of IPAN's staff have even been allowed to return to the camp from time to time and help with his care. However, at Mudumalai and within the Indian wildlife and conservation community, suspicion and resentment remain over IPAN's methods and approach. To a large extent the problem lies in the confrontational tone of their press releases that continue to accuse the Tamil Nadu Forest Department of "ecocide" and recently declared an antipoaching project at Mudumalai to be "a joke."

There is no doubt that the makhna's suffering was cause for concern. Though he may have killed a dozen people, that does not justify inflicting pain or punishment. Nevertheless, the capture and restraint of a wild animal—particularly one as large and dangerous as a bull elephant—is never going to be easy or completely free of injuries. Traditional methods used by mahouts to subdue and train an elephant are likely to offend some observers but it must be recognized that cruelty is not the motive. At the same time, the larger issue of captivity and the interaction between human beings and animals needs to be considered. As we continue to destroy the elephant's habitat and thereby force these animals into greater conflict with farmers, more situations like this will occur. Questions of response and responsibility need to be resolved, particularly the role of nongovernmental organizations (NGOs).

The forest department of each state in India has a mandate to protect forest lands and wildlife. Extensive rules and regulations are laid down by the government to implement and enforce this authority. Though the system may seem unwieldy and bureaucratic—perhaps even indifferent and insensitive—these are state institutions that have been established through a democratic process. At the same time, there is always room for change, and many NGOs in India, from the Bombay Natural History Society to the World Wildlife Fund and the Wildlife Trust of India, are actively working with forest departments and influencing government policies related to wildlife management.

In a situation such as the makhna's capture, there was obviously no time to wait for new laws and legislation. Here was an injured animal who appeared to be suffering abuse. Any sensitive person would try to respond. The problem, however, was that IPAN's strategy was to immediately internationalize the issue before a solution could be found. Rather than working with institutions and animal rights advocates within India, who would have sympathized with their concerns, IPAN directed their complaints abroad to a U.S. congressman. In addition, they exaggerated many of their claims and resorted to personal attacks on individuals. The hysterical language of IPAN's reports, posted on their Web site, does not enhance their credibility.

What is perhaps most telling about the case is the possessiveness of IPAN's demands. Not only did they voice concern and outrage over the elephant's treatment, they seemed to believe that only they were entitled to provide the makhna with care. Rather than continuing to work with forest officials and mahouts to ease the elephant's suffering, they insisted that he should be handed over to the field director of the IPAN Animal Refuge. Together with the patronizing rhetoric of their press releases, a disregard for India's laws, and the involvement of a U.S. congressman, IPAN appeared to be exploiting the situation in an effort to take custody of the elephant. Under the circumstances, accusations of cultural arrogance and neocolonialism seem justified. Though many international institutions that support the protection of wildlife on a global basis are sensitive to issues of national sovereignty, there seems to be a fringe movement of extreme conservationists and animal rights activists, mostly from the West, who feel they can ignore the laws and policies of countries other than their own. This action is done in the name of protecting endangered animals, but it reveals an effort to control and lay claim to threatened species and their dwindling habitat. Aside from being illegal, this approach does nothing to help wild or captive animals and alienates the authorities who might otherwise welcome international assistance.

The Villain Veerappan

While the controversy over Theppakkadu grabbed headlines for awhile, a much more serious and long-standing threat to elephants persists in the vicinity of Mudumalai. A gang of poachers, led by a bandit named Veerappan, live in the forests and foothills along the Tamil Nadu and Karnataka border, moving freely back and forth between the two states. For the past twenty-five years they have operated with impunity and are responsible for the deaths of hundreds of elephants as well as dozens of tigers and other animals. They have also smuggled vast quantities of sandalwood from reserve forests in the Nilgiris. In addition to these crimes, Veerappan stands accused of 119 murders, including the deaths of 32 police officers and 10 forest department officials. His biographer, Sunaad Raghuram, describes him as "India's most wanted man."

Ironically, the same forests of the Nilgiri Biosphere Reserve that provide habitat for the elephant serve as a refuge for these poachers and bandits. Though Veerappan began his career in sandalwood and ivory poaching, his criminal activities soon expanded to extortion and kidnapping. One of the most bizarre episodes in Raghuram's book, *Veerappan: The Untold Story,* relates to a group of wildlife enthusiasts whom the bandit took hostage in Bandipur Sanctuary in 1997. After capturing them, Veerappan "confiscated" several books and magazines from the rest house where they were staying. These included a copy of *National Geographic,* which had a picture of an African elephant on the cover. Carrying it off with them into the forest, along with their captives, the gang is reported to have spent hours poring over the photographs and estimating the weight of the elephants' tusks, as if the magazine were a wholesale catalogue.

On July 30, 2000, Veerappan committed his most brazen act by taking hostage Karnataka's favorite film star, Rajkumar. The seventy-year-old actor was abducted at gunpoint from his home and held for six months in the forest while the state governments of Tamil Nadu and Karnataka seemed impotent to respond. Eventually, after protracted negotiations, Rajkumar was released and Veerappan's reputation for invulnerability was enhanced. More recently, he kidnapped another victim, a politician from Karnataka named H. Nagappa, who was held captive for 106 days before his bullet-riddled body was found in the forest. In many ways Veerappan has become a modern-day Ravana, the demon king from the *Ramayana* who uproots the tusks of elephants and revels in violence and abduction. He himself would probably prefer to be compared to one of the tribal chiefs celebrated in Tamil Sangam poetry: "The lofty hill of the Kolli range that is under the rule of Ori with the strong bow, where the famished families of hilly parts full of kandal flowers, sustain themselves by selling the tusks of the fiery-eyed elephant."

During the Rajkumar kidnapping, Veerappan allied himself with elements of the Tamil separatist movement and attempted to project the image of a freedom fighter, supporting the rights of the oppressed underclasses. Born in 1952, Veerappan was uprooted with his family from their ancestral home by the construction of a dam on the Cauvery River. Along with other tribal people they were resettled in a remote village within the forests of southern Karnataka. Because of this experience Veerappan expresses bitterness toward the ruling Kannada elite. He also has a complete familiarity with the mountainous jungles where he and his gang remain hidden.

Photographs depict Veerappan dressed in forest green, with a bristling moustache and brandishing a rifle. Wearing bandoliers and glaring at the camera, he looks every bit the part of a cinema villain. Veerappan and his gang are well armed, and over the years they have successfully fought back and inflicted severe casualties during encounters with the police. Many of the tribal people of this region support him out of fear and a resentment toward the forest department with which they are often at odds. Large sections of Mudumalai and Bandipur Sanctuaries are closed because of Veerappan's activities, and he continues to take a toll on elephants and other animals.

It seems incomprehensible that a notorious criminal like this can continue to operate in the forests of modern India, avoiding arrest and prosecution. There are rumors that some politicians have been protecting him, either out of fear for their own lives or because he claims to represent a constituency they support. Whatever the truth may be, Veerappan remains a vicious predator—not a Robin Hood, as some would like to portray him—for he seems to have no regard for human life, threatening not only the elephants but those who struggle to protect them.

Temple Tuskers

Traveling farther south, beyond the Nilgiri hills, I crossed the state border from Tamil Nadu into Kerala, where I soon began to see elephants along the highway. These were tame animals, all of them tuskers, accompanied by two or three handlers. Though elephants are occasionally seen walking at the side of the road in other parts of India, they are most common in Kerala because of their ceremonial role in religious processions. Many of the major temples maintain elephant stables, from which the tuskers are hired out to nearby towns and villages. In this way the elephants generate revenue for the temple and provide an auspicious presence at festivals. Between these ritual engagements, the elephants and their mahouts travel the highways and back roads of the state.

Guruvayur Temple is one of the largest in Kerala. This shrine is famous for its Anna Kotta, or "elephant house," a walled compound in which fifty-seven elephants are kept. Most of these animals have been donated by wealthy devotees and they are fed and cared for through offerings made at the temple. Though the main shrine at Guruvayur is only accessible to upper-caste Hindus, the Anna Kotta is open to all and it has become something of a tourist attraction. The compound, once a family estate, was purchased by the temple in 1975. It is an enclosure of ten acres in the middle of the town, Punnathur Kotta. Aside from a few buildings and stables, there are two large tanks in which the elephants are bathed. Many of the animals are tethered under trees or in open sheds that look like airplane hangars.

Virtually all of the elephants at Guruvayur are males, most of them tuskers. The temple owns three or four cows, but these are segregated because there is a belief in Kerala that elephants breeding in captivity bring bad luck. The Anna Kotta is highly organized, with a superintendent of livestock and a resident veterinary doctor, as well as teams of mahouts and handlers who look after the individual tuskers.

At the gate of the Anna Kotta, one of the mahouts introduced himself. His name was Sathyapalan, a vigorous man of about forty who spoke English. He has been employed at Guruvayur for thirteen years, and like many of the mahouts he lives in a dormitory on one side of the compound. Sathyapalan's home is a village near Paravur, 150 kilometers to the south, where his wife and children live.

"For a mahout the elephant comes first and family second," he told me. "A mahout has to be there for the elephant every day, while other responsibilities can wait."

When I asked Sathyapalan if his father had been a mahout he said, "No, but he was a contractor for elephants. Now my son wants to study computers. He is not interested in taking care of elephants. It is a good occupation but dangerous work. I have broken ribs and there is a steel pin in my ankle." These injuries were caused by an elephant in musth who tried to crush him several years ago.

Leading me along the path that circles through the Anna Kotta, Sathyapalan pointed out a tusker named Padmanabhan, who had a brass chain with a polished nameplate around his neck. His broad forehead and the upper section of his trunk were freckled pink and gray. Standing patiently beneath a palm tree, he looked like a granite statue except for the gentle flapping of his ears.

"Padmanabhan is the most beautiful elephant at Guruvayur," Sathyapalan told me. "He is sixty-one years old. Last year the priests from many temples came to bid on him for their festivals and he fetched 75,000 rupees for a single night." Swaying from foot to foot, the tusker seemed to nod his head in agreement.

"The best elephants come from Kerala. They have the correct shape and the thickest tusks. Their heads are broad, with two prominent bumps at the top, and their trunks are the longest of all. Elephants from North India are usually taller but they are not as beautiful."

Farther on Sathyapalan showed me the oldest elephant in the Anna Kotta, named Lakshman. He is seventy-three and "retired," no longer taking part in religious processions. Lakshman's molars have worn down to a point where he has difficulty chewing, and he receives a "special tonic" with his meals. Pointing out his longer legs and leaner body, Sathyapalan told me that Lakshman was a North Indian elephant, from Bihar. Though a foot taller than Padmanabhan, he had a recognizably different shape, what the *Matangalila* describes as mriga or "deerlike" characteristics, as against the kumera or "regal" physique of an elephant like Padmanabhan. Sathyapalan did not use these terms; instead he identified the differences according to the regions from which the elephants came, pointing out a third kind of elephant he said was characteristic of Assam. This tusker's name was Vinayaka, and he was somewhat squatter than the other two and not so "beautiful" by Sathyapalan's standards.

The elephants that attract the highest prices for festivals are carefully groomed by the mahouts. Their tusks are polished and shaped, some of the ivory being trimmed every couple of years so that they remain tapered and symmetrical. Each day the temple elephants are bathed and their hides are scrubbed with coconut husks and pumice stones. Their toenails are filed and ointments are applied to sores and fungal infections. Their diet is regulated, and the mahouts are supervised by temple authorities, with rules and restrictions on the loads the elephants can bear and the maximum distance that the tuskers can walk in a day when travelling outside the Anna Kotta.

Sathyapalan's elephant, Junior Lakshman, is a makhna from the north, who was in musth. With all four feet in chains, Junior Lakshman looked like a shackled prisoner standing in the dock. Though we kept our distance, I could see the dark fluid oozing from his temples, as if black grease had been smeared on his cheeks. Swaying from foot to foot, he unfurled his trunk accusingly in our direction.

"He is angry with me," said Sathyapalan, laughing. "If I were to go near him now, he would kill me."

There is a distinct smell that comes from an elephant in musth. Though poets describe it as a sweet perfume—sometimes compared to the fragrance of a bride's hair—it is more of a sour, oily odor. Bees are supposed to be attracted to the flow of musth because it is like nectar, but all that I could see were swarms of flies hovering around Junior Lakshman's head. Sathyapalan told me that the first sign of an elephant going into musth is the smell that a mahout learns to recognize.

"Working with the elephant, my own body gets the same smell," he said. "No matter how often you bathe, it doesn't wash off."

At least a dozen elephants in the Anna Kotta were in musth and each of them looked miserable. Their coloring appeared darker because the mahouts were unable to give them proper baths and simply hosed them off from a safe distance. In addition to the musth fluids streaming down the elephants' faces, their penises were constantly dribbling and the insides of their thighs were lathered with urine. Piles of dung and the rotting remains of fodder surrounded each animal. From time to time they splashed water on themselves from their drinking troughs, and the ground at their feet had become a fetid mire because the mahouts were afraid to clean near the aggressive animals.

Sathyapalan explained that musth can last anywhere from a month to ninety days. "During the first stage," he said, "the elephants let us go near them and we massage their temples to release the fluids. They often get erections and using their trunks, some of them will masturbate—like humans."

Later, as musth progresses, the elephant's penis remains partially tumescent and drips urine steadily. In the wild a musth bull achieves some relief, either through mating or by rutting in the mud, but captive males are forced to endure not only their pent-up sexual tensions but the increased restrictions imposed by their captors.

Compared to the Theppakkadu Elephant Camp, the Anna Kotta at Guruvayur is a depressing place. Instead of being able to swim in a river or wander into the forest to feed, the temple elephants are confined to their enclosure and each of them is chained except when being bathed. Resting against many of the tusker's necks were long spears that are used to control the elephants if they cause trouble. The concentration of male elephants also adds to the feeling of incarceration. Whereas the captive animals at Theppakkadu approximate the social makeup of a natural herd, with females and calves, the situation at Guruvayur is anything but normal.

The danger posed by musth elephants is very real and often

underestimated. During the festival season in the winter of 2002–2003, three different tuskers in Kerala killed their mahouts and rampaged through the streets. These incidents were blamed on inexperienced handlers who forced their elephants to take part in religious processions even though they were coming into musth. When the mahouts tried to beat the aggressive animals into submission, the elephants turned on them and took revenge.

Wealthy devotees continue to donate elephants to the Guruvayur Anna Kotta. Sathyapalan explained that before an elephant is given to the temple, there is a screening process. Certificates are required from the forest department to prove it has not been obtained illegally, as well as health clearance and other bureaucratic formalities that can take several months. According to Sathyapalan, seven more elephants were waiting to be "remitted" to the temple.

The most recent arrival was a young tusker, eight years old, donated by Jayalalitha, the former film star who is chief minister of Tamil Nadu. Following a corruption scandal, in which she was eventually acquitted, Jayalalitha presented the elephant as a gesture of piety to the Guruvayur temple. Sathyapalan pointed out this elephant, who was tethered directly behind the offices of the supervisor of livestock. Restless and agitated, the young bull had a pile of palm fronds to feed on. His tusks were short and sharp, his eyes mistrustful.

Before I left, Sathyapalan took me to his quarters, a cramped and airless room he shared with two other mahouts. Though elephant handlers in Kerala have their own labor unions and are somewhat better off than mahouts in other parts of India, handling elephants is not an easy life. Sathyapalan was eager to show me the artificial tusks with which he outfits Junior Lakshman. During festivals these false teeth are attached to short incisors hidden beneath the makhna's upper lip. The imitation tusks are made of wood and painted white to look like ivory. Metal clamps at the root end screw securely into place. Holding the huge dentures in my hands, I was struck by how little they weighed.

"We have several makhnas in the Anna Kotta and also a few elephants that have lost one tusk because of accidents or infections. For festivals we attach these false teeth and nobody knows the difference," said Sathyapalan with a mischievous grin.

Though the hoax seems innocent enough, it underscores a prevalent obsession with tuskers and ivory. On my way out of the Anna Kotta, I stopped to take another look at Junior Lakshman, who was chained in the mud, angry and incontinent, his penis leaking urine and his cheeks streaked with the secretions of his temporal glands. I found it difficult to reconcile this image with the romantic metaphors of musth elephants in Tamil and Sanskrit poetry. These chained creatures cannot roam across the mountains like their ancestors. The primal instincts of regeneration, the struggle to procreate and further their species, are frustrated by the celibacy imposed on them at the Anna Kotta, and the false tusks seemed emblematic of their emasculated desires.

Facing the southern gate of the Guruvayur temple is a life-sized statue of a famous tusker named Kesavan. Though he died in 1976, this elephant is still remembered with admiration and affection throughout Kerala. On the anniversary of his death, the Anna Kotta elephants are taken in procession to the temple to pay their respects. Soon after Kesavan died, a feature film in Malayalam was made about his life, and he has achieved mythical status as the most loyal and morally righteous tusker that has ever lived.

In the movie, *Guruvayur Kesavan,* the weaknesses, fears, and failings of human beings provide a contrast to the stoic nobility of the elephant. Toddy-drinking timber contractors, well-meaning but feckless mahouts, and wicked moneylenders make up some of the cast of characters. Kesavan patiently accepts his duties as a working elephant, hauling logs at a timber yard, but in the end he revolts against the exploitation and greed of the villains. There seems to be a Marxist message in this film, as the docile, laboring elephant finally rebels against the injustices of society. In a frightening sequence, Kesavan goes on a rampage, running through the streets and trumpeting loudly as people scatter in all directions. Though Kesavan never injured anyone, he was sometimes referred to as "Lunatic Kesavan" because of these fits of madness, yet he always returned to the Guruvayur temple where he served for fifty-four years. In 1973 he was crowned "Gajarajan Kesavan" ("Kesavan the Elephant King") by the temple authorities and carried the gold emblem of the deity on his back. He even had a secret lover, Gajarani Lakshmi, one of the female elephants at the Anna Kotta. According to temple lore, "the sensual advances made by her to the majestic Kesavan and their hidden love were well known," though the relationship remained unconsummated because of temple injunctions against elephants breeding in captivity.

For the human characters in the film there is plenty of romance. The heroine is serenaded by her lover as she bathes in a pond full of lotus blossoms. Emerging from the water in a clinging wet sari, she offers one of the pink flowers to Kesavan, who watches over the two lovers like a benevolent chaperon. The elephant takes the lotus from her hand and raises his trunk in a gesture that recalls the Gajalakshmi image of fertility.

When Kesavan returns home after a rampage, he is weak and close to death. His attendants surround him and try to ease his pain but the long-suffering tusker stands motionless, without complaint. When the time comes for him to die, he sees a vision of the Guruvayur deity surrounded by a radiant light. The god appears as a young boy with peacock feathers in his hair, growing gradually in size until he dwarfs the elephant. Tears come to Kesavan's eyes as he collapses on the ground.

The final scene in the movie shows crowds gathering around

the elephant's body, showering it with wreaths, garlands, and handfuls of flower petals until the mountainous corpse is buried in floral tributes. Each of the characters is shown weeping with remorse, and even the villains lurk guiltily behind the other mourners. Not only Hindus offer their final respects to the great elephant but also Christians and Muslims. Even in death, Kesavan commands the adulation of all those who surround him, but as the camera circles his flower-laden corpse, it pauses for a moment as we catch a glimpse of his massive feet, still wrapped in chains.

Cardamom Hills

To reach the town of Thekkady, I drove for several hours up a winding hill road, ascending through plantations of coconut, coffee, rubber, and tea, as if each increment in altitude was marked by a different cash crop. Thekkady is situated at about five thousand feet above sea level in the Cardamom Hills of Kerala. The town itself is a busy cross-hatching of streets lined with hotels, restaurants, and curio shops that cater to tourists. Periyar Tiger Reserve, one of the oldest and best known wildlife sanctuaries in India, lies only a kilometer from the center of Thekkady. The approach to the gate of the park is hemmed in by enormous luxury hotels that look as if they belong in a major city like Trivandarum or Chennai, rather than on the boundaries of a protected forest.

As in many of India's national parks, there is a large reservoir at the center of Periyar, formed by dams that are part of a hydroelectric project. Most visitors view wildlife from a flotilla of boats that leave the park headquarters in the morning and in the evening. On my first day at Periyar, I bought a ticket on a boat that held sixty tourists and looked as if it might sink under our collective weight.

Even before we left the dock, I saw a pack of wild dogs, known as dhole, that were following a herd of sambar along the shore. The deer were wary and stayed together while the dogs were clearly trying to separate one of the fawns so they could run it down. Dhole are slightly larger than jackals and have reddish coats and bushy tails. In many ways they look very much like ordinary street dogs in India, though quite a bit healthier and better fed.

Heavy jungle surrounds the reservoir, with rolling hills on either side. The only disconcerting sight is the bare trunks of dead trees rising out of the water, a stark reminder of the forests inundated by the dams. With eight diesel-powered boats zigzagging through narrow estuaries, I didn't expect to see much wildlife, but we came upon several sounders of wild boar, three herds of gaur, a fair number of sambar, and two otters that were playing in the shallows of the reservoir, oblivious of the cameras that clicked in time with their antics. We also saw four wild elephants, who stood patiently near the water's edge as the boats rumbled by, about a hundred meters from shore. Three of the elephants were adult cows and the fourth was a calf that must have been five or six years old. Periyar has a population of about 1,100 wild elephants, though the percentage of tuskers to females is approximately 1 to 60, the lowest ratio in all of India. (Some estimates place the figure at 1 to 100.) Poaching has always been a problem at Periyar and continues despite forest department patrols and a global ban on the ivory trade.

Remembering the fifty-two tuskers I had seen two days earlier at the Guruvayur Anna Kotta, I couldn't help but wish that some of those captive bulls, particularly the ones in musth, might be released into Periyar's forests to help procreate their own kind. If only a furlough system could be arranged by the temple authorities and park officials.

The forests of the Cardamom Hills, along the border between Kerala and Tamil Nadu, are the ancestral home of many indigenous tribes. Most of these people once lived as hunters and gatherers, collecting forest produce to feed and support themselves. In many cases, the traditional methods of hunting and gathering by which these tribal people survived are now illegal, and their forest skills have been exploited by ivory smugglers who have turned them into poachers. Throughout India, tribal populations have been forced to make the transition from hunting and gathering to a more modern way of life, largely because their forest homes have disappeared. Despite government efforts to integrate and uplift these communities, the majority of tribals remain landless and live in poverty. They also continue to face stigmas and prejudice from upper-caste Hindus.

One of the smallest tribal communities are the Kadar nomads, with a total population of little more than 2,000, scattered through the mountains of Kerala and Tamil Nadu. A few of them live near Thekkady and subsist on collecting honey, wild spices, herbs, and aromatic resins from the jungle. The Kadar people have their own language and a distinct culture that is tied to the forest. Zacharius Thundy, an anthropologist, has studied the Kadar and recorded their songs and legends. He relates a story that describes the close relationship between the Kadar and the elephant.

A young man of the Kadar tribe married a beautiful orphan girl and she soon became pregnant. The husband immediately noticed that his bride's appetite had increased and she continually asked him for strange kinds of food. As her pregnancy progressed, she insisted that he take her into the forest and collect quantities of leaves and herbs, including palm fronds, which she devoured. One day, when the husband returned after cutting a huge load of fodder for his wife, he discovered that she had given birth to an elephant calf. Terrified, he ran back to his village and told the others in his tribe. The Kadar elders came and confirmed what had happened.

As Thundy writes: "From that day forward the Kadar look upon the elephant as a member of their own kin. Neither do they hurt elephants nor are the elephants supposed to do any harm to the Kadar. Also, it is a custom of Kadar women to refrain from over-indulging in food during pregnancy lest they give birth to monsters."

Another Kadar folktale is the story of a royal tusker who belonged to the king of Talinchiyam. One day the elephant stopped eating. Thinking he was sick, his mahout offered him all kinds of herbs and delicacies, but the elephant refused to touch any food. Eventually, an expert was consulted and he determined that the tusker would only eat if a beautiful woman was brought in front of him. The king's daughter was immediately summoned and as soon as the elephant saw her, he regained his appetite. After this, the princess and the elephant grew attached to each other. She often spent time in the stables and learned to ride the tusker. One day, while she was perched on his back, the elephant ran off into the forest, carrying the princess with him. After they were deep in the heart of the jungle, the elephant allowed the princess to alight but he refused to let her return home. Even when he went to sleep, the tusker made sure that the princess lay down beside him, resting against his flank. She was finally rescued by a prince named Thachayyappan, who killed the elephant. In recognition of his bravery he was given the princess's hand in marriage and inherited the kingdom.

Prime Habitat

Unlike the regulations at many other parks and sanctuaries in India, visitors to Periyar can enter the jungle on foot, accompanied by a "tribal tracker-cum-guide." This is part of a project organized by the forest department and an ecodevelopment society to encourage indigenous people to participate in conservation. It provides employment and is advertised as a switch "from poaching to protection."

At half past seven in the morning, I met my guide near the gate of the park. He introduced himself as Kannan, a young man in his early twenties, wearing a dark green uniform. We were also joined by an Englishman, about six foot four, with long hair that fell in ringlets to his shoulders and tattoos on both arms—lightning bolts and an ॐ. He told me that he worked as a gardener in London and was visiting India on holiday.

Kannan set off at an energetic pace, straight up a forested ridge, leaving us breathless before we reached the top. The altitude at Periyar is deceptive, for the hills don't appear that high, even though it is almost six thousand feet above sea level. Initially, the forest we passed through was mostly teak, with scattered rosewood and amla trees. Almost immediately we came upon piles of dung and teak saplings that had been stripped of their bark. Some of the branches of the trees had been torn down by elephants, which had been feeding on leaves, the green stems twisted and frayed.

One of the metaphors of courtship used in classical Tamil poetry is the image of a tusker reaching up with his trunk and breaking fresh leaves to feed his mate. These gestures of affection also have erotic connotations, as one poet writes: "The pleasure that my lover derived from me resembles the branch which was broken by the elephant but which hung on through its fibers to the tree without falling to the ground." Seeing the ravished limbs of the teaks, I was reminded of the powerful strength of an elephant's trunk as well as its amorous dexterity.

After half an hour of walking we came to a deep trench, about six feet across, which Kannan told us was the boundary of the park and the border between Kerala and Tamil Nadu. We followed this trench for a ways and passed through a patch of wild cardamom plants. Kannan broke open one of the fleshy green pods and gave us the aromatic seeds to taste—a sweet, sharp flavor that lingers on the tongue. A little farther on he

pointed out a shikakai tree, the beans of which are used for making soap and shampoo. These looked a bit like tamarind and
when broken open, they had a pleasant perfume that left a scent
on my fingers for hours.

Crossing another ridge, we made our way through thickets
of tall grass, some of which had been cut and tied up in sheaves.
A short while later we passed a group of six women with sickles
in their hands. Kannan explained that the grass they cut was used
as thatch for the roofs of their homes, in a village on the outskirts
of Thekkady. These women were fascinated by the Englishman's
tattoos and pointed at him as we walked by, as if he were some
strange, exotic species.

Our walk lasted for three-and-a-half hours that morning.
Though we saw no elephants and only a few deer, the forest itself was spectacular. The trails we followed passed back and
forth between moist-deciduous jungle and what Kannan referred
to as "semievergreen," much more lush and humid. The bird life
was as varied as the flora, with a constant chorus of calls—the
metallic chiming of drongos, the stammering cries of woodpeck-
ers, and the chuckling of mynahs. We also heard the booming of
a Nilgiri wood pigeon that sounded like the thrumming of a
double bass, though much louder and more resonant. The most
beautiful bird of all was a paradise flycatcher that flew up in front
of us. The male has white plumage with a black head. It is about
the size of a sparrow, except for the tail feathers, which are three
times the length of its body, a pure white streamer that flows behind the bird as it darts through the air. For a few seconds the
flycatcher perched on a branch overhead, then swooped out of
sight, its tail stitching the shadows together with delicate ribbons of white.

Walking through the Periyar forest was like a drug-induced hal-
lucination in which every one of my five senses was aroused. Immediately I became an addict who had to go back for more. That

same afternoon found me standing at the gate of the park with another guide named Aruvi. He was about thirty, a few years older than Kannan, but also a Paliyan tribal whose grandparents had been forest dwellers. Aruvi explained that his father had settled in the village of Kumli, adjacent to Thekkady town. Being a guide in Periyar allows men like Kannan and Aruvi to make a modest living and at the same time remain connected to their forest heritage.

That afternoon we followed a different route, through lowlands bordering a muddy stream. On either side were the remains of farms that had been abandoned with the creation of the park, guava orchards gone wild and pepper vines growing up the trunks of trees. At one place stood a large jackfruit tree, which Aruvi said was a favorite delicacy for elephants. High up on a branch, just out of reach of an elephant's trunk, hung one of the large green fruits, the shape of a lopsided basketball with a nubbled skin. As we passed from deciduous forest into semi-evergreen, Aruvi recited both the Malayalam and English names of the trees. He also knew a few of the Latin names, though this led to some confusion. One of the species he kept pronouncing as "C. G. M. Kumani," which to my ear sounded more like the name of a Tamil politician than a tree. The name baffled me until I was able to consult a book and realized that it was *Syzygium cumini,* a fruit tree commonly known as jamun.

The size of some of the trees was overwhelming. Rising over a hundred feet, they towered above the lesser species like leafy mammoths, supported by roots extending in huge buttresses that flanked the lower trunks. Many of these were a species known as jungli dungy *(Tetrameles nudiflora)*, which has silver-gray bark. Some of the trees were fruiting, and high above us I could see Malabar giant squirrels, with rusty red fur and luxuriant tails. They leaped from branch to branch as they fed and their acrobatics reminded me of otters, except that they moved through the air instead of water. We also came upon a number of black

Nilgiri langurs that make a low hooting call as a territorial signal. Their alarm cry is a higher-pitched bark, which I heard whenever we passed beneath them. The langur's fur is jet-black, except for a pale ruff around its face that makes it look as if it were wearing a mask. The coloring of these monkeys is almost the reverse of that of the Hanuman langur, common in northern India, which has white fur and a black face. The other two kinds of monkeys in the park are bonnet macaques that lurk about the boat dock, begging food from tourists, and lion-tailed macaques that remain hidden in core areas of the park.

Reentering the moist-deciduous jungle, Aruvi pointed out two sandalwood trees, twenty feet in height. Each of them was probably worth several thousand rupees. We also passed beneath enormous silk cotton trees that rivaled the jungli dungy in size, with spreading crowns. These were flowering, and the branches were full of mynahs and tree pies, as well as langurs, all of whom were feeding on the rubbery red blossoms. At one point, Aruvi scrambled up the side of the hill to a small, straight tree that grew about forty feet from the path. With a rock he began hitting the trunk of this tree, making a metallic, ringing sound. Returning a few minutes later, Aruvi showed me a handful of crystallized sap that he had chipped off the tree. When he put a match to the sap, it melted and gave off black smoke that smelled like frankincense. He said the tree is called kala dammar and it is often used for incense.

These are the forests that once covered the highlands of southern India, where elephants found no shortage of edible leaves and grasses, with vegetation so lush and dense it gives the word *biomass* new meaning. As it began to grow dark, Aruvi and I circled back toward the gate of the park. Along the way we passed a herd of sambar that let us approach within fifty feet before they went crashing off into the underbrush. Though I saw much less wildlife on our trek than I would have spotted from one of the boats, it was infinitely more satisfying. Just as we crossed the last ridge, I heard a sudden trumpeting sound, high-pitched but resonant. Repeated twice, the call seemed to come from only a few hundred meters away, though by now it was too dark for us to investigate. Aruvi smiled and nodded. He didn't have to tell me what it was.

Determined to see an elephant while on foot, I arranged to take another trek the next morning, even though my legs were sore from seven hours of walking and there were leech bites on my ankles. The guides at Periyar are assigned in rotation, and this time I was accompanied by a man named Rajappan, who was closer to my age and much less talkative than Kannan or Aruvi. He was also suffering from a bad cold and I could tell he wasn't looking forward to our walk.

Just after dawn we headed down to a shallow inlet, where a woman on a bamboo raft ferried us across for a couple of rupees. After that we didn't meet another human being for the next four hours. Along the shore of the inlet the grass was wet with dew, and I could see where animals had come down to the water's edge to drink, leaving dark trails where the grass had been disturbed. Soon we entered the forest and the air was cool but still, as if compressed by the weight of surrounding leaves. This morning I had agreed to wear canvas leggings as protection against the tiny leeches that are almost invisible until engorged with blood. Fortunately, because of the season, there were few mosquitoes in the forest. Leading the way, Rajappan waved a stick ahead of him to break the spiderwebs in our path. He remained silent and stopped from time to time, listening for sounds before proceeding.

Once again we crossed back and forth from semievergreen to moist-deciduous forest and as the hills grew steeper, the contrast increased. The northern slopes, which get more moisture, were like rain forests, while the southern slopes were considerably drier, with patches of tall grass and a more open growth of trees.

The ridgelines served as a boundary and where the trail followed the crest of a hill, I felt as if we were walking between two completely different landscapes. Gray jungle fowl scurried from one side of the path to the other, as if unable to make up their minds which forest to choose. Along the way we passed a termite nest dug up by a sloth bear. I could see the scratch marks of its claws in the dirt. The core of the nest was a spherical chamber that looked like a coconut shell. The bear had broken this open and scooped out the larvae inside.

After an hour and a half, we dropped down into a broad valley that was mostly grassland, ringed with groves of *Terminalia* trees. This area looked promising, though there weren't any animals in sight. We passed a water hole dug by the forest department, next to which grew a flowering bonfire tree. Its vermilion blossoms and the turquoise plumage of a kingfisher perched on a branch provided a brilliant contrast to the tawny expanse of grass. As we crossed the valley and went over a low rise, the sun was warm against our backs. Beyond us lay an inlet of the reservoir, a silvery tongue of water.

Both Rajappan and I saw the elephants at the same moment. They were swimming about fifty feet from shore, rolling over on the surface, almost like whales or porpoises. The water was not very deep and when they stood, I could see the upper half of their bodies. Soon afterward, the elephants started to wade out onto the bank, their trunks waving and their black silhouettes like fluid shadows emerging from the lake. They were a small herd of six cows and two calves, the youngest of which was probably a year old. Standing in the shallows at the edge of the reservoir, they sprayed themselves and seemed reluctant to leave the water.

We were about two hundred meters away. Between us and the elephants there was no cover, though we circled around through a patch of scrub jungle and approached within a hundred and fifty meters. It was difficult to know if the elephants had caught our scent but if they had, the herd seemed not to care, splashing about in the mud. The youngest calf was particularly playful and ran back and forth between the older elephants, as if trying to catch their attention.

For half an hour we watched the herd before they started to move off. Neither Rajappan nor I spoke as our eyes followed their profiled shapes. The colors and contrasts in the landscape were muted now that the sun was directly overhead, but the elephants, still wet from their bath, stood out sharply against a backdrop of dense forest and scalloped hills. Their movements were graceful and unrestrained. No mahouts were seated on their backs. No human commands directed them. No chains or ropes held them in place. As the elephants made their way toward the trees, they moved not as in a procession or parade. Instead they seemed to drift through the grass with absolute freedom, guided only by the shared instincts of a herd, the companionship of their own kind.

murals, monoliths, and miniatures

Cave Frescoes

*Y*ears ago wild elephants would have wandered freely through the jungles bordering the Waghora River. Feeding on the abundant foliage and bathing in clear pools, herds of elephants must have made their way up into these winding gorges, especially in summer when the surrounding landscape was scorched and brown. This remote, uncultivated valley, along the northwestern rim of the Deccan plateau in central India, was chosen by Buddhist monks in the first or second century B.C. as a spiritual retreat called Ajanta. Seeking to isolate themselves from the world, these monks excavated cave dwellings and shrines out of the smooth basalt cliffs overlooking the Waghora. The earliest monasteries were simple, austere structures, but in the latter half of the fifth century A.D., Ajanta attracted the patronage of princes and nobles in the court of King Harishena. Though Harishena was a Hindu ruler, the Buddhist shrines were greatly expanded during his reign (A.D. 462–481) and many more caves were excavated. Over a relatively short period of time, some of India's most spectacular art was created inside these cave temples, priceless vaults of carvings and exquisite frescoes.

As the elephants reached the upper end of the valley, where the river carves an oxbow in the plateau, they must have startled the monks out of their meditation. Breaking branches and splashing about in the stream, the herd would have disturbed the silence and seclusion of Ajanta. At the head of the gorge, their progress was blocked by a waterfall that plunges through channels in the rock, forming seven pools as it descends.

Looking down from the ridge above, one can easily picture the scene as if it were happening today, though wild elephants are no longer found in this part of India. The monks come scrambling out of their cave viharas, squinting in the bright sunlight and pointing at the herd, waving their arms and shouting to chase them away from fruit trees and gardens along the water's edge. The animals are equally surprised to find human beings living in this gorge. Unable to continue upriver, they turn abruptly and retreat downstream, trumpeting in frustration and stripping leaves from vines and branches.

The walls of the caves were plastered with layers of mud and limewash, then painted with mineral pigments such as red and yellow ochre, kaolin, gypsum, and cinnabar. Scenes from the Jataka tales, which relate allegorical episodes in the Buddha's life and earlier incarnations, decorate the walls of the caves. Though the monks of Ajanta lived in the Spartan darkness of tiny chambers and slept on stone beds, they reconstructed the world they had renounced and projected these images onto the shadowy interiors of their shrines. The frescoes illustrate stories of Prince Siddhartha in sensuous detail—a vivid contrast to the ascetic isolation and natural surroundings of the Waghora gorge. Hidden within the caves are scenes of royal pageantry and lovemaking, haughty figures with strings of gold jewelry and elaborate hairstyles, as well as courtesans and celestial Apsaras, nymphs that would have tempted even the most celibate imagination. Altogether twenty-six different caves were excavated at Ajanta, but just as quickly as this surge of artistic activity began, it ceased, at the end of the fifth century. The temples and monasteries at Ajanta were abandoned and remained virtually forgotten for over a thousand years. Once again, the elephants would have been able to wander upriver, undisturbed and unaware of the treasures hidden within the caves.

On April 28, 1819, an English cavalry officer named John Smith was hunting along the Waghora River. Entering the gorge in search of wild boar or tiger—depending on the guidebook's embellishments—he spotted a carved archway in the cliff, partly obscured by creepers. Exploring farther, Smith came upon several of the temples where he found the frescoes and sculptures, which had remained in darkness since A.D. 500. Soon after Smith carved his name and regiment inside a cave, Ajanta was recognized as one of the most important religious and artistic sites in the world.

While the walls of the cave temples are painted with narrative images from the Jataka tales, frescoes on the ceilings are purely decorative. Lotus blossoms and other floral motifs hover overhead, a richly embroidered canopy of colors. The tour guide's dim flashlight picks out each flower and bird as it emerged from the brushes of the ancient artists, working patiently in the flickering aura of oil lamps.

The delicate designs and patterns on the ceiling reflect an aesthetic sensibility rooted in nature—images of waterbirds and deer, leaf patterns, and the coiled shapes of vines and flowers. As the guide leads me through the cave, I stumble on the uneven floor but keep my eyes fixed upon the ceiling and the jungle above me. Once again, I am searching for an elephant, and there it is in a square panel of its own—a white elephant in a playful pose, its trunk curled around a lotus blossom. I have seen this image many times before; it has been adopted as a logo by India's Tourism Department. But here in this cave, with the surrounding panels of flowers and foliage, the elephant seems almost alive— "sporting," as the poets put it. The strokes of the artist's brush are visible against the peeling plaster. My guide moves the beam of his flashlight about, tracing the elephant's shape. For a moment it feels as if I am outside the cave and the animal is there in front of me. I can see it as the painter may have done, his eyes blinded by sunlight after hours in the dark. Perhaps his work has been inter-rupted by shouts and clapping, his meditative art disturbed by a commotion near the river. Stepping out of the cave onto a narrow ledge, he sees a young elephant in the river below him, playfully tossing weeds with its trunk. Making no sound to chase it away, the artist watches every movement of the animal's body, memo-rizing its shape. Later, returning to the cave, he deftly sketches an elephant on the ceiling with a brush dipped in cinnabar, then paints the image that I see above me.

Jataka Tales

Patil, my guide at Ajanta, had a master's degree in English litera-ture. He was a dignified, scholarly man, who introduced me to one of the other senior guides by saying, "I am a Hindu. He is a Muslim. But we often sit together and talk about Shakespeare." Though the cave temples at Ajanta are Buddhist monuments, Patil discussed the history and theology of the monks who created these structures as if their traditions were his own. Showing me a book that he was reading on Buddhist art, he said, "I am always learning more about Ajanta, studying the latest publications."

When I explained that I was searching for elephants, Patil immediately pointed out a carved frieze above one of the cave entrances. It depicted a sequence of elephants—two tuskers fight-ing, others walking in procession, one killing a tiger, and another trampling a cobra. Farther on, at the entrance to a different cave, two large elephants were carved into the face of the cliff, each of them with a bent foreleg, kneeling in respect for the Buddha. This image recalls one of the Jataka tales in which the Buddha subdued a rampaging elephant called Nalagiri. When the tusker, who had trampled many people and was enraged with musth, saw Gautama, he immediately knelt down and touched the Buddha's feet with his trunk, then threw dust onto his own head.

Inside the caves, Patil led me to a number of frescoes that contained elephants. The first of these related the story of the Buddha's birth, the familiar image of his mother Maya dreaming of a white elephant entering her womb on the night her son was conceived. The queen is shown in her bedchamber, surrounded by attendants as the elephant appears above her. Paint and plaster have flecked away from the wall so that it was difficult to see, but Patil directed my eye with his torch. Restoration of the frescoes is an ongoing task and sections of the walls were being cleaned and stabilized by specialists from the Archeological Survey of India. Ajanta has been identified as a World Heritage Site by UNESCO, which helps fund the restoration. Considerable damage has occurred since the caves were rediscovered, and photography is prohibited to protect the delicate colors in the paintings from the caustic glare of flashbulbs.

Instead of being separated into distinct panels, the murals at Ajanta flow into each other, so that the narrative is continuous. The rich earth colors of the paintings have a warmth and brilliance that make them seem to glow in the feeble torchlight. Beyond the dreaming queen sit the court astrologers who interpret her vision of the white elephant. They recognize it as a sign of royalty, the vehicle of Indra, and an omen of great significance. The astrologers predict that the child will become a powerful emperor or a great saint. As the story unfolds across the wall, we see the pregnant queen in Lumbini Forest, where the infant Siddhartha is born and miraculously takes his first steps.

Moving from image to image, Patil recounted each story in a voice that conveyed its emotional timbre. Unlike the recitations of most guides, his telling did not seem rehearsed, though he must have related these tales hundreds of times. Later, as he began to show me other Jatakas containing elephants, Patil stopped himself and apologized. "I can tell you all of the facts and explain the pictures but expressing the feelings is very difficult," he said. "It isn't often that I tell these stories."

More familiar to him were accounts of the Buddha renouncing his kingdom and leaving his beautiful wife, Sundari. Or the subtle imagery surrounding his first sermon in Benares, where each person in the crowd bears a different expression on his face, ranging from awe to arrogance. These were the stories that Patil knew by heart, though as he began finding more elephants for me—shaking his torch whenever the batteries began to fail—and outlining lost sections of the frescoes, which had crumbled off the walls, he warmed to his subject and found his voice.

The Gaja Jataka, also known as the Matiposaka Jataka, is an allegory that resembles King Dasharatha's story in the *Ramayana*. The Bodhisattva (Buddha-in-waiting) is incarnated as a wild elephant who has two blind parents. He is a devoted son and leads them through the mountains and forests, guiding the elderly couple to lush stands of bamboo and pools of water where they can drink and bathe. Eventually, this dutiful elephant is captured by a forester who takes him to Benares, where he is tied up in the king's palace. Overcome with anxiety for the fate of his parents, the elephant refuses to eat, no matter what his captors offer him. The king realizes that something is wrong and he sets the elephant free. Running off into the forest, the tusker is reunited with his mother and father. In a final scene, he drapes lotus blossoms on his parents' heads while the blind mother wraps her trunk around her son's leg in a gesture of recognition and affection.

We carry on through the caves to another Jataka that Patil finds with his Eveready torch. Again the Buddha takes the form of an elephant, but this time there is a terrible drought and many people are starving. In an act of supreme sacrifice, the elephant throws himself off a cliff and dies, so that human beings can feed on his flesh. The fresco depicts the starving throng, their faces contorted with hunger, looking up at the elephant who stands serenely at the top of a cliff. We also see his corpse lying on its side, the Bodhisattva having given up his life so that others might live.

Chadantta Jataka, the third of the frescoes that Patil showed me, was the most elaborate. Here again he hesitated, saying it was difficult for him to express the nuances of this story. In the first scene, a tusker is shown bathing with two female elephants in a pond covered with lotus blossoms. This is the king of the elephants, known as Chadantta, for he has six tusks. His consorts frolic on either side of him, their trunks sensuously splashing the water. The elephant king takes a lotus blossom and offers it to the younger consort. Overcome with jealousy, the other cow is so distraught she leaves the pool and dies of envy. But as with everything in Buddhist mythology, there is a cyclical pattern, and the jealous elephant is reborn as the Queen of Benares. Recalling the bitter experience from her past life, she tells the king where he can find the elephant with six tusks and asks that he bring her the ivory. Hunters are quickly dispatched and track down Chadantta, who patiently awaits his fate. Being a Bodhisattva, he already knows the purpose of their mission. Wounded by a hunter's arrow, he wraps his trunk around each of his tusks and uproots them one by one, after which he dies. When the hunters arrive at the palace in Benares, bearing bloodstained ivory on their shoulders, the jealous queen realizes what a terrible thing she has done and collapses with guilt and grief.

The Stature of Mountains

While the temples and monasteries at Ajanta are exclusively Buddhist monuments, the cave complex at Ellora, a hundred kilometers to the south, contains Buddhist, Jain, and Hindu shrines. Each of these faiths found inspiration in the basalt escarpments along the margins of the Deccan plateau. Boring into these ridges, artisans quarried the precious ore of their religious heritage—deities and myths, symbols and metaphors chiseled out of solid rock. To enter the cave temples at Ellora is to feel the power of negative space, the absent weight of all the stone that has been removed, the hollow cavities surrounding the pillars and idols that remain.

Both the *Ramayana* and the *Mahabharata* epics relate the story of King Sagara's sixty thousand sons, who dug deep into the earth while searching for a royal stallion. When they reached the foundations of the world, instead of a horse they discovered the primal elephants that stand at the cardinal points of creation. In the eastern quarter they came upon the noble Virupaksha, who stood patiently with the land and forests resting upon his back. In the southern quarter stood Mahapadma, as large as the mountains that he supported. In the western quarter the sons of Sagara discovered Saumanasa, to whom they offered their obeisance, and to the north stood Himpandura, bearing the weight of the Himalayas on his shoulders. These elephants are the cornerstones of the cosmos that have stood there since the world was formed.

Like King Sagara's sons, the armies of artisans who cut their way into the basalt ridges at Ellora excavated elephants at the foundations of these shrines. The tiered Kailash Temple, the most dramatic of all the rock-hewn monuments at Ellora, is supported by the backs of hundreds of elephant caryatids. Though many of these elephants have been damaged, they retain a stoic dignity that has survived the centuries since they were carved. Unlike the animated images of gods and goddesses that decorate the temple walls—depicted with multiple arms and shown firing arrows and trampling demons—the elephants have a noble permanence, as if carrying the weight of all that mythology, the tumult and the towering conflicts of good and evil. Mount Kailash is the snow-covered peak where Shiva presides over the cosmos, and the temple re-creates the geography of his domain. The elephants themselves rise up like the Himalayan ridges that buttress Kailash. At the base of the temple, in a shadowy, subterranean world, is the demon Ravana with his ten heads and twenty arms, who fights with the elephants and shakes the mountains.

On either side of the Kailash temple stand two life-sized monoliths of elephants, draped with carved garlands and bells but unencumbered by the weight of the cave structures. These ceremonial figures stand apart from the temple itself, on separate pedestals, dignified and motionless, perpetual votaries offering homage to the Hindu gods. Like the celebrated tuskers at Guruvayur Temple in Kerala, they add grandeur and gravity to the rituals performed within this sanctuary.

The first deity to be worshipped at Kailash is Ganesha, who occupies his traditional place at the threshold of the shrine. But the image that catches the eye immediately upon entering the temple is of the goddess Lakshmi sitting on a lotus. Two tuskers raise their trunks above the goddess and shower her with water. Gajalakshmi—Lakshmi of the elephants—is an icon of fertility, one of the most fundamental images of Hindu tradition. The goddess emerges out of a lotus, containing the seeds of life within her, a vision of beauty and perfection, her breasts swollen, her gestures inviting, her face set in a benign attitude of motherhood. And the elephants, holding water vessels in their trunks, bring down rain upon her, life-giving showers that fill the carved pool of lily pads at the base of the idol.

After seeing the frescoes at Ajanta, in which the Buddha's mother dreams of the white elephant floating above her like a cloud and then descending into her womb, it is difficult not to understand the connection between these two myths of fertility. Gajalakshmi represents a distillation of elephant lore, the tuskers as water bearers, as clouds. In this scene the fecundity of the goddess and the life forces that emanate from her body are generated by the libations of elephants. Not only do they stand as pillars of the temple, its stolid foundations, they are also an active and essential part of creation. It is the water that pours from their trunks that rejuvenates the earth and ensures the regeneration of life.

———

Elephants can be found in the earliest examples of Indian art, long before the sculptures and frescoes at Ajanta and Ellora were created. Terra-cotta seals from the Indus Valley civilization at Harappa and Mohenjodaro display the easily recognizable profiles of tuskers, some of which appear to be adorned with decorations or tied with ropes, evidence that elephants were captured by man as early as 2,000 B.C. The seals also reveal hints of animistic traditions, in which the elephants were probably revered as icons of nature. Though an inch in height and simply rendered, these faience relics remind us of the enduring presence of *Elephas maximus,* part of a tradition and culture that existed long before there was any distinction between Hindu, Buddhist, Jain, Muslim, or Christian. Tiny clay tuskers present their silhouettes beneath undeciphered characters of a forgotten language, yet we cannot fail to understand what these images mean. They are repeated again and again in temples all across the subcontinent, from Taxila to Konarak, from Amravati to Sanchi and Barhut, from the ruined palaces at Hampi to Rajput forts at Kotah, Bundi, and Jaipur.

Tuskers are found not only in classical Indian iconography but also in folk art, like the dokra figurines from Orissa that are molded out of brass, or woven basketry images from Bihar, made of sikki grass. Elephants of all shapes and sizes appear in clay, wood, stone, and textiles, from hand-block-printed cotton to embroidered silk. They are drawn on the mud walls of village homes and stenciled on the sides of rickshaws, buses, tractors, and trucks. In paintings from Madhubani, in the northern state of Bihar, rural women artists have created stylized elephant forms to illustrate their myths and legends.

On the southeastern coast of Tamil Nadu, at Mamallapuram, lie beach temples constructed by Pallava kings in the seventh century A.D. Here again the elephants appear, and this time they are much larger than the terra-cotta seals from the Indus Valley, almost life-sized. Carved out of boulders rather than molded out of sun-baked clay, these elephants are neither restrained nor adorned.

They do not carry the weight of Mount Kailash on their backs, though they are surrounded by a host of celestial deities, many of them flying through the sky above the elephants. In this tableau, it seems as if all of Hindu mythology has gathered together, figures from Vedic hymns, Puranic lore, and epic narratives. But the largest and most imposing figures of all are the elephants—a tusker and his mate, accompanied by calves that play between their legs.

The rock carvings at Mamallapuram depict the descent of Ganga, the Ganges River, which falls from heaven to earth. This is a critical moment in Hindu mythology, as important as the churning of the primordial sea. Not only does it represent the fertility that the river goddess Ganga brings to the land but also the purity of her celestial waters that wash away the residue of death and bear the soul from one life to the next. In the carved boulder at Mamallapuram is a central channel that conforms to a natural fissure in the rock, through which rainwater pours whenever there is a storm. The gathered host of deities, demigods, mystics, and animals surround this channel. The elephants, too, stand patiently, awaiting Ganga's descent. As in the Gajalakshmi image, these are the attendants of the goddess, her water bearers and harbingers of the monsoon.

Hathinama: A Mughal Portfolio

Among the most famous images of elephants in Indian art are miniature paintings that illustrate the memoirs of Mughal emperors from the sixteenth and seventeenth centuries. Though the elephants are reduced in size, they dominate the scenes in which they appear. The emperors who rode upon their backs recognized the elephant's stature as a projection of their own ambitious designs. Babur, the first Mughal emperor, was a Turkish chieftain who conquered most of northern India by 1526. An entire chapter of his memoirs, the *Baburnama,* is devoted to a description of Indian animals and birds, beginning with the elephant. An accompanying miniature shows a wild herd that could easily be the same family carved into the rocks at Mamallapuram. A tusker, two females, and a calf stand by the shore of a lake. One of the cows is feeding on bamboo, while the bull raises his trunk to test the air. The calf darts forward to the water's edge, chasing off a pair of ducks. Despite this idyllic scene, Babur's primary interest lay in the use of elephants in warfare. Following his decisive victory in the Battle of Panipat and the fall of Delhi, "droves of elephants were caught and presented by the elephant keepers." Babur and his successors recognized the strategic importance of elephants and maintained large pil-khannas, or "elephant stables." His memoir lists the virtues of elephants, with only one note of reservation: "They can easily carry heavy loads across large and swift-running rivers. Three or four elephants can haul mortar carts that would take four or five hundred men to pull. They eat a lot, however: an elephant eats as much as two strings of camels."

Akbar, the grandson of Babur and the most dynamic Mughal emperor, ruled from 1560 to 1605. Among his many interests and accomplishments was riding elephants. Not content to be carried about in a ceremonial howdah, Akbar learned the methods of mahouts and prided himself on being able to control tuskers in musth, a skill that he believed was granted him by God to impress his subjects. The *Akbarnama* is full of hyperbole for the emperor as well as his chosen mount.

> When India was made illustrious by his blessed advent [Akbar] gave special attention to elephants, which are wonderful animals both in form and in ways. If in respect of size I liken them to a mountain . . . I do not succeed in my attempt. . . . Or if I liken their speed and fury to the wind, how is their wrath depicted at the time of their o'erthrowing the firm-footed on the field of battle? If I compare them in foresight, intelligence and sagacity to the horse the real thing is not said.

Unlike the Buddha, who subdued elephants with a benign gesture of his hand, Akbar aggressively dominated the tuskers in his pil-khanna by force of will. The *Akbarnama* goes on to claim that "without exaggeration he rode more than a hundred times on mast elephants which had killed their drivers and were men-slayers, and were capable of smiting a city or perturbing any army." Akbar often staged elephant fights as a form of entertainment, sometimes mounting one of the tuskers himself, "when the fumes of wrath were circulating in its brain." Akbar then provoked the tusker to attack another bull. In a dramatic miniature painting the two elephants are shown charging across a bridge of boats that are in danger of capsizing, while attendants are flung to the ground or into the river. A version of this story is recounted in the memoirs of Akbar's son, Jahangir, who quotes his father as saying:

> One day, in the full bloom of youth, having drunk two or three bowls of wine, I got on a must elephant. Although I was sober . . . I pretended that I was dead drunk and that the elephant was uncontrollable. . . . Then I called for another elephant and set the two to a bout. Fighting madly, the two elephants rushed toward the bridge over the Jumna. . . . I thought that if I checked it from going over the bridge, the people would know that my drunken actions were a pretense and it would be obvious that neither was I drunk, nor was the elephant out of control and such actions are unsuitable for kings.

Though Akbar has been lauded by historians for his liberal philosophies, he was brutal and unforgiving to those who opposed him. One particularly gruesome miniature depicts the various forms of execution administered by elephants. While one tusker prepares to trample two cowering traitors, another gores a man with its tusks. The bloodied and barely recognizable remains of two corpses lie on the ground, while a pair of elephants rip apart the remaining victims, in a scene that was clearly intended to frighten anyone who plotted against the emperor. Despite these sadistic punishments, Akbar could be sensitive to the emotions of elephants. His memoirs relate the story of a female elephant who starved herself to death after the tusker with whom she was paired had been killed in combat. Inspired by her loyalty and love, the emperor comments: "When such marks of affection appear in beasts, what deeds are there which may not be displayed among human beings? But no one must reason from this to persons who are human in shape, but not in reality, for such persons are lower than fossils."

Jahangir was a devoted naturalist and his memoirs contain detailed descriptions of animals and plants. Elephants were one of the emperor's favorite subjects and he commissioned a number of portraits of the tuskers he received as tribute from feudal chiefs. Jahangir also had a habit of granting honorific titles to his elephants while admitting that he himself was not above "poetic conceit." Once, after being eulogized in verse, he gave the poet an elephant in recognition of compliments paid. Unlike Akbar, who constantly forced his elephants into combat, Jahangir was more concerned with the welfare and treatment of his tuskers. In the *Jahangirnama,* he writes:

> Of all animals the elephant particularly likes water and loves to get into it, even during the cold weather of winter. If there is no water available, it will take water from a bag and spray it over its body. It therefore occurred to me that, no matter how much elephants enjoy water and are accustomed to it by nature, surely during winter they must be affected by the cold water. I therefore ordered the water heated to lukewarm and poured into its trunk. On previous days when it sprayed cold water over itself, the effects of shivering and trembling could be seen, whereas, in contrast, it seemed to enjoy the warm water. Such treatment is peculiarly mine.

The next two Mughal emperors, Shah Jahan and Aurangzeb, maintained their family's obsession with elephants. In a painting from the *Shah Jahan Nama,* Aurangzeb is shown on a white stallion charging straight at a tusker, into whose trunk he drives a spear. In the background is another elephant—this one a makhna—restrained by men with charkhis, or hollow lances filled with fireworks. The memoir provides commentary: "One day . . . His Majesty enjoyed the spectacle of a combat between two intoxicated royal elephants. In the course of the conflict, finding the animals were getting out of sight, he followed after them on horseback, attended by the Princes Muhammad Shah Shuja and Muhammad Aurangzeb, who rode up to the left of the elephants to witness the exciting scene. . . ." Suddenly, one of the tuskers turned and charged directly at Aurangzeb. Instead of trying to escape, the prince "galloped at it and rose up in his stirrups and hurled his spear with his whole strength against its forehead. The monster, smarting from the wound, again rushed furiously to the attack and [Aurangzeb] . . . being struck in the flanks by the elephant's trunk rolled over from the violence of the shock. The instant Prince Aurangzeb fell from his saddle to the ground, he sprang nimbly up, and stood sword in hand, ready to strike." This incident foreshadows the fearless and ruthless nature of the prince. Several years after surviving his encounter with the tusker, Aurangzeb imprisoned his father and killed his brother before ascending the throne.

While Mughal artists produced many detailed drawings of animals that have the quality and precision of scientific illustrations, they also painted pictures of fantastic beasts and demons. A striking miniature from Akbar's atelier portrays a herd of elephants caught in the clutches of a demonic rukh, believed to be the enemy of elephants. Though usually described as a giant bird, in this miniature the rukh is an enormous winged monkey with a tusker's head. It fights with a simurgh, a phoenix, that comes to the rescue of the elephants. A very similar image, possibly inspired by this painting, appears in a Mughal carpet, now in the collection of the Museum of Fine Arts in Boston. In the knotted tapestry a herd of seven elephants struggle to escape the clutches of the rukh, including one wrapped in its trunk and another in its tail. Muhammad al-Damiri's fourteenth-century bestiary *Hayat al-Hayawan* describes the rukh as laying eggs that are as large as domed buildings and "each of its wings is 10,000 fathoms long." Mentioned by Marco Polo in accounts of his eastern journeys and appearing in stories of Sinbad in the *Arabian Nights,* the rukh is said to be capable of lifting elephants in its talons, though more often it grabs huge boulders and drops them from the air onto ships. In such Mughal miniatures the elephant is transported and transformed, through human imagination, from the natural world into the realm of the supernatural.

Artifice and Nature

The *Hastividyarnava,* an ancient Assamese text of *Gajashastra,* compares the cosmos to an elephant's tusk. Human beings are said to reside at the point of the tusk, the asuras (demons) in the middle, and the devas (gods) at its root. This schematic vision presents the elephant as an infinite and eternal creature, much larger than anyone could ever imagine. The tapered universe of ivory serves as a linear diagram of divinity, malevolence, and mortality. To live at the tip of an elephant's tusk is to know the threatening void that lies beyond.

Ivory has always been an object of desire, a pure and precious material representing luxury and wealth. It has been carved into every imaginable shape, from nudes to crucifixes, from the handles of walking sticks to phallic implements of erotic design, not to mention billiard balls, piano keys, and cigarette holders. As an artistic medium it has been fashioned into animal forms, including miniature statues of the elephant itself. In sculptures like these the tusk and the tusker become one, a grotesque and disconcerting mutation of substance and subject. Those who find beauty in carved ivory see its whiteness as essential, a solid, blank mass out of which the human imagination can create whatever images it chooses. From the time of the early Greeks and Romans, the greatest demand for ivory has always come from places where the elephant does not exist. One of the objects unearthed in the ruins of Pompeii was an ivory carving of an Indian yakshi that appears to have been the handle of a mirror.

An elephant's tusk lends itself to carving because of its unique combination of strength and elasticity, as well as its creamy luster. These qualities are a result of the structure of the dentine, which is formed out of microscopic tubular grains fused together and growing gradually in length throughout an elephant's life. As with any tooth, a nerve runs through the center of the tusk. The upper portion, embedded in the socket, remains hollow and is filled with pulp. Only a third of the ivory can be used for any substantial piece of carving. The outer surface, sometimes referred to as the rind or bark, is generally removed, and the inner core is worked by craftsmen, limiting the scale and shape of the sculpture. Ivory sometimes has impurities or distortions and occasionally bullets have been found encysted within a tusk. In one instance it was discovered that a newly turned billiard ball that failed to roll true contained a lead slug fired by a hunter years before.

Indian artisans carved ivory at least as early as the third century B.C., fashioning it into religious images as well as ornaments like bracelets and combs. However, the reverence for elephants in Buddhist and Hindu culture, as well as their considerable value as tame animals, would have prevented the wholesale collection of ivory as a commodity. The Mughal emperors and their successors used ivory to decorate their thrones and for other lavish objects of art. Under British rule in India, the supply of ivory was further exploited. Large quantities of elephant tusks were also imported from colonies in East Africa, and ivory carving grew into a sizable industry in places like Kerala and Rajasthan.

Examples of ivory artifacts can be found in almost every museum in India, but they have disappeared from the curio shops and emporia where they used to be on sale only a few decades ago. Since 1976, by the Convention on International Trade in Endangered Species of Wild Fauna and Flora (CITES), the sale of ivory from Asian elephants has been banned. Trade in African ivory ended in 1989, with the exception of two controlled sales of government stockpiles. Though commerce in ivory continues illegally, many of the Indian artisans who were once employed as ivory carvers have turned their chisels and saws to other materials, like wood and horn. Today the primary market for tusks remains East Asia, specifically China and Japan, where ivory artifacts such as miniature Buddhas or hanko seals still hold a perverse value.

As an artistic motif, the elephant has been exploited in countless ways that go well beyond literal representations of the animal itself. In many Indian paintings, from the fifteenth century onward, elephants are transformed into something other than themselves, though they retain at least the outline of their physical shape. These composite paintings often depict elephants created out of entwined human forms, usually the bodies of beautiful women. Female figures are fitted together like pieces of a jigsaw puzzle that yield an elephant's profile. The thighs and legs of one woman become its trunk. The outstretched arms of another serve as its tusks. The braided hair of a third hangs down as the tail, while her back is arched to form the elephant's rump. The contortions of the women, which allow them to fit inside the tusker's body, seem to suggest a human desire to inhabit an elephant's skin. Many of these composite paintings show Krishna riding as a mahout, for he is the divine and insatiable lover who has intercourse with innumerable partners—an image that mirrors the ecstatic experience of faith and devotion. As he rides upon his elephant, made up of female forms, we are reminded of Krishna's adoration as well as a musth elephant's irrepressible desires. Kama, the god of love, is also featured in this kind of erotic imagery, shooting arrows of passion from the back of a tusker.

Later miniature paintings, beginning with the Mughal period, portray elephants constructed out of other animals and monsters. As in the painting of the rukh, several different creatures are fused together to create a metaphorical beast. Not only elephants, but also horses, camels, and tigers are formed out of a compressed menagerie of animals. One of the most intriguing of these composite paintings can be found in the Prince of Wales Museum in Mumbai. Attributed to an anonymous nineteenth-century Deccan artist, it depicts two tuskers fighting. The untitled miniature contains several human figures, though most of the images that make up the two elephants are other species of wildlife, such as tigers, lions, jackals, bears, and buffaloes. The trunk of one elephant is a snake, while the feet of both tuskers are rats. The two mahouts are demonic figures, themselves constructed out of various animals. One has the head of a stag, the other an ankush made of a fish. The entire painting presents an outlandish, hallucinatory blend of creatures compacted together to portray the terrible fury of dueling elephants.

Somewhat similar to these composite paintings are tugra designs created by calligraphers who use the fluid lettering of the Persian script to produce images of elephants. Instead of drawing human and animal forms to fill in the shapes, these artists employ words and phrases so that visual images are created entirely out of consonants and vowels. In most cases the verses or epithets have little to do with the elephant—invocations to god or praise for a king—but the form of the animal itself can be clearly seen within these flowing arabesques. The trunk and tusks, the bulbous forehead, are all described by the calligrapher's pen as he bends and twists the alphabet into the shape of a royal ele-

phant. A dotted vowel serves as its eye, while elongated consonants produce its legs. On the back of the elephant rests an ornate howdah, also constructed out of words.

In a mildly subversive manner these tugra designs circumvent
orthodox Islamic injunctions against representative art. The words
themselves become a picture and the abstract beauty of language
charms the eye with a linguistic riddle. On one level it is an elephant and on another it is nothing but words, though these have
become so convoluted and intertwined that only an expert can
actually decipher what the calligrapher has written. Like the composite paintings, tugra calligraphy redefines nature in metaphorical terms. Studying the ornate tangle of words, one cannot help
but realize that the elephant has literally become the text.

Kotah

Some of the liveliest and most expressive paintings of elephants
come from the fortified city of Kotah, in Rajasthan. Here the
seventeenth- and eighteenth-century court artists—several of
whom apprenticed under Mughal painters—transformed the elephant from an imperial icon into a creature of grace and eccen-
tricity. These paintings often focus on the playful and combative
nature of elephants. Though the majority of Kotah artists remain
anonymous, scholars have assigned titles to some of these minia-
turists in recognition of their distinctive styles. One of the artists
is known as the Master of Elephants. Art historian Stuart Cary
Welch describes him as follows:

> However well this vigorous artist rendered people, trees,
> black bucks, or cows, only his elephants were truly visionary.
> Without them, his accomplishment might be overlooked.
> Because he flowered artistically at Kotah, a major center of
> elephants, he could scrutinize his archetypal subject daily.
> And if we compare his elephants to those of virtually any

other of the world's artists, most of whom imbue them with the ponderousness of battered old sofas, it is evident that only he fully appreciated their catlike antics and the look of bones, gristle and muscle fluidly packed into thick, flexible skins. Heightening empathy he calculatedly exaggerated hollows and lumps, and contracted or expanded distances between anatomical parts.

Another unique aspect of the Master of Elephants is the way in which he depicts the confined tension in captive elephants, their restless and aggressive nature, restrained by the bonds of human masters. One of his most powerful images is a khaka sketch, an ink drawing that has been lightly colored, which bears the title *An Angry Elephant Breaks Its Chains.* The artist portrays a regal makhna, his feet wrapped in iron, though many of the links have been shattered and the elephant holds remnants of fetters in its trunk and mouth. With ears flared and tail twisted like a corkscrew, the makhna seems ready to trample anything in its path. In the lower right-hand corner of the drawing, two attendants hold up spears in a futile attempt to keep the elephant at bay.

The Master of Elephants must have spent long hours studying the anatomy and behavior of these creatures, which were housed in stables below the city palace. His patron, Rao Madho Singh—the first ruler of Kotah—was clearly proud of his elephants and must have delighted in the lifelike images created by this artist.

Above the main entrance to the city palace in Kotah, a pair of carved tuskers raise their trunks to form an arch. Known as the Hathia Pol, this brightly painted gateway admits visitors to a privileged domain of audience halls and private chambers. Within the city palace and museum, elephants are everywhere. The golden howdah in which the Maharao was borne on a tusker's back and the silver howdah for his heir are both on display, along with a life-sized statue of a tusker wearing gilt head-gear and embroidered raiments. Even the palanquins in which the royal family was carried are decorated with finials in the shape of elephant heads. A silver elephant, part of an antique table setting, bears a candelabra on its tusks, and a marble statue of Airavata, with Indra and Indrani on his back, has four tusks and seven trunks, coiling out like the tentacles of an octopus.

Many of the palace rooms are decorated with frescoes of elephants, some carrying the Maharao in procession, others depicting scenes of wild elephants frolicking in pools of water. The Chattar Mahal, one of the royal suites, contains a mural of Rao Rama Singh hunting a rhinoceros from the back of an elephant. The same shikar scene appears in a famous Kotah miniature, where the wild-eyed elephant wraps its trunk around the rhino's neck. The palace frescoes also include gentler images, one of a dancing girl turning a pirouette atop a platform supported on the tusks of an elephant. Before electric fixtures intruded on the frescoes, wall brackets in the shape of elephants held oil lamps in their trunks. Sections of the Bada Mahal, another suite of the palace, are decorated with marble friezes, carved with elephants engaged in sport and combat. Two tuskers square off like wrestlers, their trunks entwined. In another scene an elephant pins down a hapless camel. There is also a carving of the elephant-headed rukh, in which the winged demon holds a herd of struggling elephants in its trunk and talons. This image is virtually identical to the earlier miniature painting and carpet from the Mughal period.

Seated in a terraced garden overlooking the Chambal River, Maharao Brijraj Singh could easily be one of the figures in a painting from the court of his ancestors. A handsome, clean-shaven man in his late sixties, he exudes a regal dignity. On a swing beside him sits his wife, Maharani Uttara Devi, wearing a sari of indigo silk. Their son, Maharaj Kumar Ijyaraj Singh, is also present.

A servant with a yak tail fly whisk stands discreetly in the background. The scene has the composition of a formal miniature, overshadowed by the dark foliage of a kajlia tree, its leaves providing a shingled dome of green. Peacocks wander about the garden and cranes nest along the riverbank.

"Elephants used to be part of virtually all the royal ceremonies and festivals at Kotah," says Brijraj Singh, explaining how they once played an important role in the Dussera celebrations that reenact episodes from the *Ramayana.*

On the tenth and final day of the festival, the Maharao rode out of the palace gates on his royal tusker. Cannons were fired from all corners of the walled city, and most of the populace would turn out to watch the spectacle, re-creating the climax of the *Ramayana,* when Ravana, the demon king, is killed. On the Ramlila grounds an effigy of Ravana was erected, with a masonry torso, its ten heads and multiple arms made of bamboo and tissue paper. At sunset the Maharao fired an arrow at the demon, piercing a pot of water dyed red like blood. Then the tusker attacked the effigy and tore off its arms and heads.

"One year the elephant wasn't able to tear off Ravana's heads and people waited until ten o'clock at night when someone was sent up to knock it down. After that my grandfather, Maharao Umed Singh, installed a pulley system to topple the heads in case the elephant didn't succeed."

Brijraj Singh rues the fact that there are no longer any elephants in Kotah. The last of the royal tuskers was sold by his father in 1971 to Tiger Tops Jungle Lodge in Nepal. Nowadays, during Dussera, the demon's effigy is filled with fireworks and explodes in a blaze of gunpowder. "It's all over in a few seconds," complains the Maharao. "With the tusker, you got to see Ravana being ripped apart. Much more dramatic!"

During the spring festival of Holi the rulers of Kotah sprayed colored water on their subjects from the back of an elephant. A bawdy, orgiastic festival, during which all forms of licentious behavior are permitted, Holi is celebrated with intoxication and revelry, and the passions it provokes are often compared to the spirit of musth.

According to the German Sanskrit scholar Heinrich Zimmer, the *Hastyayurveda* contains an account of a similar festival devoted to the worship of elephants as symbols of fertility. Every year after the crops were planted, a royal tusker was painted white with sandalwood paste and paraded through the streets, accompanied by groups of men dressed in women's clothes, who performed lewd rituals in order to arouse the regenerative forces of nature. "If they did not pay worship to the elephant, the king and the kingdom, the army and the elephants would be doomed to perish. . . . [but if these rituals were observed] Crops will sprout in due time; Indra, the rain god will send rain in due time. . . . Whoever wishes to have sons will have sons, and longings after riches and other goods will also be fulfilled."

"During Holi everything is permitted. A man can fondle any woman he wants and there is no separation between king and subjects," says Brijraj Singh. Of course, riding on his elephant, the Maharao of Kotah had a distinct advantage, though people showered him with color from the rooftops. Several paintings in the city palace show Maharao Ram Singh, who ruled from 1827 to 1865, playing *Hathi ki Holi,* armed with a long hose attached to a fire department pump that spewed jets of color like a water cannon.

"Maharao Ram Singh loved to play Holi. He would go through the streets of the city on his elephant and spray women watching from doorways. Their dresses flew up and their bodices burst open from the force of the water," Brijraj Singh recounts with animated gestures.

Maharani Uttara Devi gives her husband a skeptical look, while Ijyaraj Singh listens to his father's stories in amusement. The family now make their home in Brijraj Bhawan, formerly the British Residency. Unlike many maharajahs, the royal family of Kotah has adjusted to the changes that overtook India in the twentieth century. They still maintain a privileged lifestyle but live within a modern world. After independence, Brijraj Singh joined politics and served as a member of Parliament until 1977. He is keenly interested in the heritage of Kotah and has published books and articles on the art and culture of his kingdom. Though Brijraj Singh does not maintain an atelier as did his ancestors, he preserves the work of earlier artists and has established the Rao Madho Singh Museum Trust that maintains collections in the city palace.

Brijraj Bhawan itself is a colonial building, with deep verandas and bougainvillea vines. Inside, the walls are hung with hunting trophies, most of the animals shot by Brijraj Singh's father, Bhim Singh II. The dining hall is lined with the heads of tigers, leopards, gaur, and deer, as well as an incongruous moose. Shikar is a fundamental part of Kotah's lore and many of the miniature paintings depict hunting scenes. Elephants were often used to beat for tigers, and Brijraj Singh recalls these hunts during the 1940s and 1950s. "I remember roads through the forest that were so densely covered by trees that we used to have to turn on the headlights of a car during the day."

Much more inspiring than the hunting trophies is a silver urn that sits in the dining room at Brijraj Bhawan and is used for washing hands. The lid is decorated with an image of the ancestral patron of the Chauhan clan, from whom the rulers of Kotah trace their lineage, a four-armed figure rising out of a fire. The spout of the silver vessel is shaped like the head of a tusker with its trunk upraised, while the tusks operate as a faucet, moving up and down, regulating the flow of water.

Even in death the elephants of Kotah attend upon their kings. Just outside the city walls lie the royal cenotaphs. Being Hindus, the deceased maharaos of Kotah were cremated and their ashes immersed in the Ganga. Though they have no tombs, their successors constructed domed chattris to commemorate each ruler's passing. These cenotaphs are located in a garden known as Kshar Bagh. Each chattri has a pair of plaster tuskers standing guard at the foot of a staircase that leads to a raised plinth supporting a stone cupola. Around the base of these memorials are carved panels in which the images of elephants predominate—two bulls in combat, a mahout restraining a royal tusker.

When I visited Kshar Bagh, the sun was just setting and the spectral shadows of mango trees surrounded the cenotaphs. Much of the garden is unkempt and the older chattris lie in ruins. Several of the plaster elephants are riderless or broken, the ground littered with debris. Mendicants live at the back of the garden, ghostly figures who watched me from a distance while their fires burned near the city wall. As the shadows grew darker, I couldn't help but feel an eerie presence. All at once, a troop of rhesus monkeys leaped down from the trees and approached me with demanding looks in their eyes, thinking that the bag in my hand contained something to eat. Their aggressive behavior was unnerving, and only after I picked up a stone to throw at them did they let me leave. As darkness fell over Kshar Bagh, the regal silhouettes of the elephants stood in silent homage to those former kings who rode upon them in festivals, in battle, and in sport. Yet these memorials seemed as much a tribute to the elephants themselves, the cenotaphs like empty howdahs, and the smoke from the mendicants' fires drifting through the trees like a vaporous herd.

VII

arundhati's bath

Ritual Ablutions

*a*t Hardwar the Ganga flows out of the Himalayas and debouches into the plains of northern India. A crowded pilgrimage site, attracting millions of devout Hindus every year, Hardwar was once a remote forest retreat favored by reclusive mendicants. Even today these sadhus carry on the ascetic traditions of Vedic sages, though instead of sitting under banyan trees in quiet contemplation they often position themselves at busy crossroads, smoking hashish and offering spiritual counsel to passersby. Over the past century, Hardwar has grown into a major town, connected to Delhi by highways, rail links, and a nearby airport. Areas of forest still stand on either side of the river, though severely degraded, a final vestige of the Terai jungles that once extended all along the base of the Himalayas from Nepal to the Punjab. The weathered range of Siwalik foothills forms a crumbling barrier between the mountains and the plains. Much of this region has been cleared for cultivation, but the forests near Hardwar are preserved as Rajaji National Park, home to approximately 400 wild elephants.

I arrived in Hardwar on June 9, 2002, which happened to be an auspicious day for immersion, the riverbanks overflowing with pilgrims taking a ritual dip in the Ganga to wash away their sins. Har-ki-Pauri—the footprint of God—is the most sacred place for bathing, where the river has been diverted through an artificial channel with cement steps leading into the water. Temples and ashrams line the shore and a clock tower rises above the bathing ghats. Though the footprint of God can be traced back to the moment of creation, at Har-ki-Pauri the hand of man is evident in every other respect. The Ganga flows between high embankments and breakwaters, its current regulated by engineers. Diversion of the river began with construction of the Upper Ganges Canal in the nineteenth century. Since then this holy stream has been rerouted and contained through an elaborate network of barrages, headworks, hydroelectric projects, canals, and bridges. As a result, the original course of the Ganga, downstream from Hardwar, is reduced to a sluggish trickle in summer. Only after the monsoon's arrival does the river actually debouch in all its unrestrained glory, reaffirming the myth of Ganga's descent from heaven.

By the second week of June, rain had yet to fall in Hardwar, though clouds were massed against the mountains and the humidity pressed down on the land with a heavy stillness. Rather than rutting against the foothills like Kalidasa's tusker, the clouds looked like a herd of elephants preparing to charge across the plains. Temperatures at noon were above 100 degrees Fahrenheit and the only respite lay in the river, its water ice-cold from the glaciers out of which the Ganga flows. Thousands of bathers were swarming over the terraced ghats—white-haired pensioners stepping gingerly into the current, children splashing about in the nude, ascetics, government servants, farmers and merchants, men and women, pilgrims from every corner of India. They lingered in the chalky current, submerging themselves completely, then rising with folded hands—refreshed, rejuvenated, rinsed clean of human guilt.

Across the river and four kilometers upstream from Hardwar another immersion was taking place. This bather was alone in the

Ganga, recumbent as the current swept past her in swirls of glittering silt. With languorous gestures, she luxuriated in the cold water, disappearing beneath the surface for several seconds, then rising slowly as rivulets cascaded off her skin. Rolling onto one side, she seemed oblivious of everything except for the fluid caresses of the Ganga. The midday sun was bright and there was no shade along the riverbank, only round white boulders and the harsh glare of sand. Farther back from the water lay the jungle, a ragged line of shesham, ber, and sal trees.

The lone bather worshipped the Ganga in her own way, not as a human supplicant but as a creature born to water. One moment she frolicked in the sacred river, splashing and spraying herself with innocent delight, then all at once she collapsed and let the stream wash over her, supine as a boulder, scoured and sculpted by the current.

Arundhati. She is named after the planet Venus and the dutiful wife of the Vedic sage Vashishta who once lived in jungles along the banks of the Ganga. But this Arundhati is an elephant who belongs to the Uttaranchal Forest Department. She lives at Chilla on the east bank of the Ganga, in Rajaji National Park. Every day her mahout, Zahur Ahmed, brings Arundhati to the river for a bath. He sits patiently on the shore, allowing her to spend as much time as she wishes in her ablutions. Like the elephants carved into the rocks at Mamallapuram, Arundhati attends upon the Ganga as it flows out of the mountains and begins its long journey to the sea.

Off in the distance the rooftops of Hardwar are visible against the folded ridges of the Siwalik Hills. The sounds of pilgrims—chanting busloads, recorded hymns blaring from loudspeakers, the cumulative babble of a hundred thousand voices—is drowned by the rustling murmur of the current. A short distance upstream from where Arundhati lolls in the river is the Chilla Hydro Project, an enormous, brutish structure of concrete and steel. But here, where the elephant bathes, there is forest on either bank and a few thatch huts where herdsmen keep their cattle.

Eventually Arundhati rises from the Ganga, glistening in the afternoon sunlight, her skin almost black. As her forelegs straighten, she leans back on her haunches and sucks a last draught of water into her trunk. Reluctant to leave the cool river, Arundhati sprays herself, then stands. Over seven feet tall, she is a full grown cow of forty. As the elephant turns to look at me, I can see the mottled pigmentation on her ears and trunk and around her neck, pale markings as delicate as the spots on a butterfly's wing. Wading from the water, she is magnificent, unencumbered and unadorned.

Zahur Ahmed, his white beard tinged orange with henna, speaks to her from the shore—a combination of commands and conversation. The elephant listens and stops at the water's edge, then lowers her head so the mahout can grasp each of her ears before scrambling up her trunk. Though Zahur is in his sixties, he remains agile and Arundhati boosts him effortlessly onto her back. They set off for home, across the riverbank, past the herdsmen's huts, and into the jungle that borders the forest complex at Chilla. Along the way Arundhati pauses several times to uproot clumps of cannabis growing wild in the forest. After beating these against her legs to dust off the roots, she swallows enough marijuana in each bite to keep a sadhu drugged for weeks. The cannabis appears to have no effect on her.

The Language of Elephants

My first encounter with Arundhati occurred six months earlier, in late December. On a winter morning, with a dense mist rising off the Ganga, I waited for Zahur Ahmed to take me for a ride in the park to look for wildlife from Arundhati's back. In contrast to the sweltering heat of June, it was close to freezing and I was bundled up in a sweater and coat, blowing into my cupped hands for warmth.

Arundhati came lumbering out of the mist with a canvas pad on her back, as if she had just risen from bed and was still draped in a quilt. After I climbed aboard, we headed across a dry river-bed and into the forest near Chilla. From time to time Zahur Ahmed allowed Arundhati to stop and feed. Coiling her trunk around one of the lower branches of a haldu tree, she began to tear off the heart-shaped leaves.

"Not those ones, Arundhati," said Zahur. "Take the leaves up above. *Dalle Upar.*"

Obediently she released the branch and reached higher into the tree where the foliage was thicker and the stems were easier to break. From the beginning I noticed that Zahur spoke to his elephant in a gentle, chiding manner, accompanied by com-mands. Farther on, we came to a rotting tree, partially eaten by termites, and he demonstrated how Arundhati could push it over with her forehead.

"Lagey!" said Zahur, urging her on with his knees pressed be-hind her ears. Leaning against the trunk, she slowly pushed and without much effort uprooted the dying tree. The crash startled a pair of sambar deer, who darted out in front of us. Fumbling with my camera to try and get a picture, I accidentally dropped the lens cap, which disappeared in the thick undergrowth and layers of dead leaves on the ground. Seeing what had happened, Zahur muttered another command, *"Dhat Dalle."*

Immediately, Arundhati halted and turning around she cast about with her trunk among the leaves. Within thirty seconds she had located the lens cap, picked it up with her trunk, and handed it to Zahur. Her eyesight may have been poor, but her sense of smell was so acute that she had no trouble finding the small plastic disk. The dexterity of her prehensile finger made it possible for her to pluck my lens cap from the ground. No other creature has an appendage as versatile as an elephant's trunk, which is sensitive enough to sniff the faintest scent, capacious enough to suck up several bucketfuls of water, sensuous enough to engage in foreplay, and strong enough to kill a man with a single blow.

As we carried on through the forest for a couple of hours, I was struck by the level of communication between Zahur and Arundhati. Every time we saw a deer or wild boar, he whispered a command and she would stop as if he'd pressed a brake. Instead of using an ankush, Zahur carried a short bamboo stick, though I never saw him strike Arundhati. His commands were accompanied by affectionate banter and a nudge of his knees or toes behind her ears. If Arundhati was slow to respond, he would raise his voice sternly but never in anger. Occasionally, she seemed to answer him with a fluttering sound that elephants make when communicating with each other, a low-pitched burbling noise that came from deep inside her lungs and made her flanks vibrate like kettledrums. Whenever Arundhati approached another of the forest department elephants, a smaller cow named Chanchal, the two of them would make this sound, as if carrying on a contented conversation. Katy Payne, an American zoologist, has demonstrated that elephants communicate with each other using low-frequency calls, most of which are inaudible to the human ear. In her book *Silent Thunder* she describes listening to a recording made of Asian elephants in a zoo. "With the tape running ten times its usual speed, we heard the calls, condensed and nearly three octaves too high—a little like the mooing of cows. . . . Two animals had been carrying on an extensive and animated conversation below the range of human hearing."

As for communication between elephants and man, a good deal of speculation surrounds the origin of commands used by mahouts, and theories have been put forward about this so-called elephant language. Some believe it is derived from Assamese and others claim that the commands are a pidgin dialect of Persian and Urdu, dating back to the time of the Mughal Empire. Words from different languages can be found within this vocabulary, but as Zahur Ahmed explained, "It depends on what an elephant is taught. Each can have its own commands." The romantic notion that a common language exists between man and elephant is an exaggeration. However, mahouts in various parts of India employ a remarkably similar set of commands, even in places where the regional languages are completely different. The continuity of these commands points to a tradition passed down from one generation of elephant keepers to the next. When I compiled a selection of words and phrases that Zahur used with Arundhati, these were virtually identical to a glossary that appears in A. J. W. Milroy's *Management of Elephants in Captivity,* published in 1922.

Zahur Ahmed comes from a family of mahouts. Most of his life has been spent in the company of elephants, beginning as a charrawalla, or "fodder cutter," and gradually working his way up to becoming the senior mahout at Chilla. Though he is solely responsible for Arundhati's care, including her feeding and bathing, Zahur has two charrawallas to assist him in cutting leaves and taking the elephant into the jungle to graze. Every few months a veterinarian examines her health and treats any sores or infections she might have, though Zahur expresses scorn for these doctors and says that he looks after her health, using traditional herbal remedies. He proudly told me that his elephant had never needed an injection.

The communication between elephant and mahout, after twenty-two years in each other's company, goes well beyond a simple list of commands. As I watch them together, they seem to have a relationship that is neither exploitative nor sentimental. Each understands the other and though Zahur exerts control over his elephant, he also appreciates that she is physically much stronger than he. Arundhati can be irritable at times and he is able to gauge her temper. One evening, while we were sitting outside Zahur's house, directly opposite Arundhati's stable, she was feeding on a pile of leaves. After some time she had stripped off whatever she could eat and all that was left were bare branches and twigs. To get our attention Arundhati began flinging the sticks at us with her trunk. For the first time I heard Zahur shout at her, *"Birey!"* She stopped immediately but watched him intently until he turned to one of the charrawallas and told him to replenish her pile of fodder.

Despite their quarrels, this elephant and her mahout seemed inseparable. Arundhati expressed her affection by touching Zahur with her trunk, just as elephants caress each other in a herd. For his part, Zahur was not overly demonstrative but whenever he spoke to Arundhati, his voice conveyed a paternal fondness. It was only when he told me how he became Arundhati's mahout, however, that I sensed the depth of their relationship. Zahur recounted the story without embellishment or emotion, yet he made it clear that she meant more to him than just his livelihood.

"Originally, Arundhati belonged to a bania [merchant] in Pilibhit, who kept her as a status symbol, to show off his wealth. When the bania died, his three sons divided up his property and all of his money. None of them wanted the elephant so they gave her to their widowed mother. That was all the poor woman got from her selfish and ungrateful sons. For awhile the bania's widow tried to look after Arundhati as best she could, but an elephant is expensive to maintain and her sons had left her with nothing else. There was an old mahout who tended the elephant and took her into the jungle to graze, but after awhile the forest department refused permission for cutting fodder. Finally, in desperation, the old widow offered to sell Arundhati to the forest department for 20,000 rupees. Nowadays an elephant costs ten times that much.

"After she was purchased, I was assigned to bring her here, all the way from Pilibhit. We walked together through the forest for over a hundred and fifty kilometers. It took us several weeks because I had never ridden Arundhati before, though she soon learned to obey my commands. Ever since then we have been together; for twenty-two years she has been my responsibility."

Terai

The Siwalik Hills that run through the center of Rajaji National Park have eroded over centuries into unusual shapes, strangely similar to the termite castles that protrude from the forest floor. These crenellated ridges of sedimentary rock and spires of clay are riven with dry water channels, landslides, and labyrinthine gullies. The deciduous forests that cover their slopes are much more diverse than the plantations of sal trees, teak, and eucalyptus propagated by the forest department in other areas of the park. Some of India's richest fossil deposits have been found in the Siwaliks, including the remains of prehistoric elephants like *Stegodon ganesa,* whose tusks extended ten feet beyond their sockets.

Today the descendants of those ancient creatures find scarce refuge in the Siwaliks, as the erosion of their habitat continues. Rajaji Park extends for 820 square kilometers, but this area has been fragmented by roads, railway lines, and canals that cut through the forest and impede the free movement of animals. The growing towns of Hardwar, Rishikesh, and Dehradun extend right up to the boundaries of the park, which is also surrounded by cultivation. Poaching is a serious problem, and over the past two years several tuskers have been slaughtered in Rajaji, possibly by the same gang that has been killing elephants in Corbett Park.

Five hundred years ago Mughal emperors came here to hunt the one-horned rhinoceros, but like the fabled creature that shares its Latin name, *Rhinoceros unicornis* has long since disappeared from Rajaji. The park contains no more than fifteen tigers and perhaps two or three times that number of leopards. Other animals in Rajaji range from jackals to the largest antelope in India, known as the nilgai. The number of wild elephants remains relatively stable, approximately 400, though this population is unable to roam freely and often comes in conflict with human beings living within or near the perimeter of the forest.

The presence of Gujjar herdsmen in Rajaji Park is just one example of how environmental issues in India are inevitably human issues as well. Tribal people like the Gujjars have lived in the Terai for generations, coexisting with wild animals and sharing the resources of the jungle. But population growth within these communities and the rapidly shrinking and degraded conditions in the forest have skewed the balance, as cattle compete with elephants and other wildlife for water and fodder. A large part of the problem lies outside the boundaries of the park. Traditionally these nomadic graziers spent six months in the Terai and six months in the Himalayas, herding their buffaloes up to meadows over 8,000 feet above sea level. This meant that lowland forests in places like Rajaji used to be virtually empty of cattle and human beings from April to October.

With decreasing access to Himalayan pastures, the Gujjars have come into conflict with other herdsmen in the mountains. The traditional routes they once followed are also affected by new motor roads that make it impossible for Gujjars to migrate as they earlier did. Many now remain in Rajaji and other parts of the Terai for twelve months of the year. This means that the forest, already under pressure from expanding cultivation and other encroachments, is given no respite. As fodder becomes scarcer and scarcer, Gujjars are forced to lop the branches of trees to feed their cattle. Naturally, they build their huts near water holes. This means that wild animals in the park have fewer places where they can come to drink in summer. For elephants, who require plenty of water, this has become as serious a problem as the destruction of fodder trees and grasses. Not only do the buffaloes deplete water sources, they pollute them with dung and urine. These cattle also carry diseases such as anthrax, which is fatal to the elephants.

None of this is meant to suggest that the Gujjars are to blame. As a minority community without land of their own, they face extreme challenges in a rapidly changing world. Their relationship with the forest department has often been adversarial, and it is difficult to establish trust as solutions are broached. Just as the elephants have faced a shrinking habitat, so have these nomadic herdsmen. As a result, they themselves have often been the victims of elephant attacks. One Gujjar that I spoke with described how he delivers milk to Hardwar by bicycle and only a few months earlier he had been charged by a tusker along the forest road. Though he was able to save himself, the elephant destroyed both his milk cans and his bicycle, trampling it into a heap of mangled spokes and twisted handlebars.

Efforts have been made to relocate the Gujjar population but land is scarce and resources are limited. In 1983 the government initiated a resettlement program at a place called Pathri, south of Hardwar. A number of Gujjar families moved there but many more still remain within the park. Clearly, the herdsmen themselves must be engaged in the process of finding solutions, and incentives for them to leave must be increased.

Political change has also had an impact on Rajaji Park, offering possibilities for improvement but also raising many concerns. In November 2000, the new state of Uttaranchal was created out of hill districts and sections of the Terai that were once part of Uttar Pradesh. This includes the watershed of the Ganga, along with Rajaji and Corbett National Parks. After the state boundaries were redrawn, responsibility for wildlife conservation shifted to Uttaranchal's forest department, staffed by officials reassigned from Uttar Pradesh. Though national parks receive funding from the central government, the state still bears a significant part of the financial burden. The creation of Uttaranchal was intended to give the people of this region more control over natural resources, but one of the difficulties has been limited revenue. The state government is struggling to pay for some of the most basic infrastructure and services, like roads, water, and electricity. Ecotourism is a growing industry within Uttaranchal and there is potential for attracting more visitors to places like Rajaji, but in the short run, wildlife conservation is not a priority of politicians, since human beings cast votes, not animals.

Unfortunately, the problems facing Rajaji Park require urgent and expensive solutions. Efforts are being made by government agencies and nongovernmental organizations to improve conditions at Rajaji, but much more money is needed. The Wildlife Institute of India, a training and research facility founded in 1982, has its campus on the outskirts of Dehradun, adjacent to the park. One of the senior professors, A. J. T. Johnsingh, has studied elephants in Rajaji for many years. He uses the word *compression* to describe the circumstances under which these animals exist.

"As more and more encroachment occurs, the elephants come into greater conflict with human beings," he says. "One of the biggest problems with Rajaji is that there is no core area, where the animals are undisturbed. They are constantly under stress."

Johnsingh and his colleagues have published articles promoting the idea of linking Rajaji and Corbett National Parks by maintaining and replanting corridors of reserve forest in the Terai. Their research has shown that a few male elephants have been able to travel from one forest range to the other but most herds are confined to limited territories. He believes that forest corridors would allow for the migration of elephants between these sanctuaries and promote healthier breeding and a more viable genetic mix. Even within Rajaji, Johnsingh and others have been fighting to provide wildlife with a safe passage between the eastern and western banks of the Ganga on the outskirts of Hardwar.

The construction of canals and waterworks at the Chilla Hydro Project has contributed to the problem. Though elephants are excellent swimmers and can easily cross the river, the steep embankments of the canals are impossible for them to negotiate. A narrow footbridge, constructed over a canal below Chilla, is occasionally used by bull elephants but it is not large enough for herds to cross over. Adding to this fragmentation is the main highway between Hardwar, Dehradun, and Rishikesh, which has become increasingly congested, so that animals can only move back and forth after midnight, when traffic eases. Still other impediments exist. A crucial crossing point near the village of Raiwala is blocked by an army ammunition dump. Another stretch of forest land through which elephants once passed was recently allotted to farmers who have been resettled because of the Tehri Dam, seventy kilometers upstream from Hardwar.

When one hears Johnsingh describe the problems, they seem insurmountable, but he is a man who doesn't appear easily discouraged. Though he works within the bureaucracy of wildlife conservation, attending conferences, writing papers, and trying to influence policy makers, Johnsingh is clearly a naturalist at heart. He grows suddenly animated when describing one of the largest tuskers in Rajaji, whom he named "Big Boss."

"We studied him for five years and observed him many times, but he never charged at us. A magnificent bull, over three meters tall and probably weighing five tons," said Johnsingh. "Sadly, he was killed in a fight with a younger rival, gored to death. They were both following a female in estrus and Big Boss was in musth. The bulls fought for hours, chasing each other through the jungle. Forest guards who witnessed the battle said they could hear the trumpeting from far away. In the end, Big Boss slipped and fell into a ravine and the younger elephant killed him with his tusks."

In the magazine *Sanctuary,* Johnsingh has written an elegy for this elephant: "Thus ended the life of a magnificent and noble bull that ruled the Chilla forests for more than two decades. For his imposing size, he had magnanimously allowed us to approach him very close on several occasions. Though we are meant to be dispassionate scientists, we did feel a sense of loss."

Hidden within a thicket of lantana, near Khansrao Forest Rest House, is a memorial to another elephant. This gravestone, erected by Major Stanley Skinner over seventy-five years ago, dates back to a time when sections of Rajaji Park were maintained as shooting blocks by the forest department. Hunters like Skinner were able to reserve Khansrao for several weeks of shikar during the winter season. In the nineteenth century Skinner's ancestors—part of the Anglo-Indian aristocracy—owned large sections of the Terai. The family patriarch, Colonel James Skinner, was known as Sikander Sahib, a flamboyant soldier of fortune who raised the first cavalry regiment in the Indian Army. Unlike his famous ancestor, Stanley Skinner seems to have preferred elephants to horses and the gravestone offers a touching tribute:

TO THE MEMORY OF
MAJOR STANLEY SKINNER'S
DEAR ELEPHANT
RAMPYARI
DIED HERE ON THE 6TH AUGUST, 1922
HAVING GIVEN HIM THE FINEST SHOOTING
IN THIS WORLD, FOR 14 YEARS
NEVER TO BE FORGOTTEN.
BRAVE AS A LION, STEADY AS
THIS ROCK.

It is still too early to write an epitaph for the wild elephants in Rajaji. Though they have been forced into a smaller and smaller area, with less food and water, there is still a chance that sufficient habitat within the park can be protected and preserved for their survival. Rampyari's bones have joined the skeletal remains of prehistoric Stegodons that once ruled the Terai, and tuskers like Big Boss continue to fall victim to younger rivals.

This is part of the inevitable cycle of nature, but human beings are accelerating the process of extinction. Hopefully this crisis can be averted so that in the future other elephants like Arundhati and her wild relatives can continue to bathe in the Ganga.

Uncertain Arithmetic

My visit to Chilla in June coincided with the dates for an elephant census in Uttaranchal. Unfortunately, the census also overlapped with auspicious bathing days at Hardwar. This confluence of events provided a dramatic illustration of the competing forces at play in Rajaji Park. Because of the huge influx of pilgrims, police closed the main highway between Hardwar, Rishikesh, and Dehradun, diverting all of the traffic along a forest road that runs past Chilla. This meant that a constant stream of buses, trucks, and cars were passing through a section of the park.

With the highway closed, the forest department had an opportunity to earn revenue by charging toll. Long lines of vehicles queued up on either side of the traffic barrier at Chilla. While forest officials scribbled receipts, shuffled carbon copies, and made change, the drivers blasted their horns and revved their engines impatiently. Over the course of three days, the entrance to the park at Chilla was like a busy intersection during rush hour. Even in the middle of the night, the stillness of the forest was disturbed by the pandemonium of engines, horns, and abusive voices.

At the center of this chaos stood the range officer, Jaggat Singh (not his real name), who suddenly found himself master of all that moved between Hardwar and Rishikesh. He is a man of about fifty, short-statured except when he sits astride his Enfield Bullet motorcycle, which elevates him slightly and adds a pulsing beat to his approach. Jaggat Singh presents an image of commanding indifference. His moustache is dyed orange with henna, but the white roots have grown out over a line of crooked teeth. The ranger's eyes are heavy-lidded, with a perpetually bored expression, and his reluctant smile changes to a scowl in a fraction of a second. He is the kind of government official who believes that he was appointed to his position with the sole purpose of obstructing whatever may cross his path.

At this particular moment it was a line of vehicles extending half a kilometer in either direction. Parking his motorcycle and seating himself in a plastic chair under the shade of a yellow laburnum tree, Jaggat Singh supervised the toll collection with a combination of satisfaction and distaste. All of the forest department personnel, including foresters, guards, clerks, and even the caretaker of the rest house, had been pressed into service— lowering and raising the boom, writing receipts, collecting money, and directing traffic.

When one of the truck drivers tried to jump the queue, edging out of line and advancing toward the checkpoint down the wrong side of the road, Singh barked at the guards to lower the boom and bring his cane. Rising angrily from his seat, he swaggered forward and whacked the bamboo lathi against the grille and fender of the truck, as if it were a recalcitrant elephant that needed to be brought into line. The truck driver began to protest loudly but as the beating continued, he finally shifted into reverse.

Like many of the other state institutions in Uttaranchal, the forest department is desperately short of funds, and three days of toll collection provided far more revenue than entrance fees to the park, tickets for elephant rides, and charges levied at the rest house. In fact, the first day I was at Chilla the forest guards were so busy collecting toll that nobody was free to accompany me into the forest, a requirement for all visitors. When I asked about the elephant census, Jaggat Singh nodded but gestured toward the herd of vehicles, as if to say he couldn't neglect the duty of collecting toll. During the three days I was at Chilla, I saw no evidence of an elephant census being undertaken. Neither did I see any wild elephants, for they had wisely retreated into the deepest recesses of the jungle, far away from the onslaught of traffic.

Afterward the chief wildlife warden released figures for the elephant population in Uttaranchal, announcing that there were a total of 1,500 wild elephants in the state. This would seem to be an optimistic number, though it is probably closer to the truth than figures in earlier years, which were grossly inflated. A census of this kind will always involve a certain amount of ambiguity because of the elusive nature of elephants and their considerable range. Obtaining an accurate census is a fundamental problem and as Vivek Menon points out in his book *Tusker,* "There is no bigger controversy in wildlife conservation than the numbers game." Population figures are used to emphasize the threat of extinction but their accuracy is difficult to gauge. With animals like tigers, pugmarks are often used to generate figures for a census, but even though the pads on their feet can be as distinctive as fingerprints, identifying the differences leads to many errors. With elephants, one method of conducting a census is to measure the quantity of dung in a forest and thereby mathematically arrive at a total number of animals. But again, the variables are so pronounced that the method seems almost useless. In the end, population figures for wild animals like the elephant will always be subject to discrepancies and human error.

Within the forest department there are many committed and conscientious officials, but there are also a number of individuals like Jaggat Singh, whose priorities have become warped and jaded. When funds are low, morale is bound to suffer. It is much easier to sit in the shade by the side of the road and collect toll, rather than tramping about the forest in search of wildlife. There may well have been other forest officials assigned to conduct the census, whom I didn't meet, but clearly the staff at Chilla were preoccupied with manning the toll barrier.

Another example of the problems facing Uttaranchal's forest department became evident on my first night at Chilla. The electricity in the rest house where I was staying had been off all day and as the temperature and humidity climbed, it became unbearable indoors without a fan. Even though all of the windows were open, the air was stagnant except for the feverish whirring of mosquitos' wings. Soon after nightfall I noticed many of the other buildings in the forest complex had electric bulbs that were burning brightly. When I approached Jaggat Singh, he looked at me with an indifferent expression.

"Actually," he said, "the problem is that the forest department hasn't been able to pay its electricity bills and 160,000 rupees are owed. That's why the connection to the rest house has been cut off."

He paused a moment and then continued, "But if you pay the linesman fifty rupees, he'll connect the wires illegally."

Setting aside any ethical qualms—no one is ever blameless in these transactions—I handed over a bribe and the caretaker of the rest house was sent off in search of the linesman. Within half an hour the electricity came on and the ceiling fans began to stir the air. Though it is easy enough to find fault with the forest department, the problem is much greater than one or two individuals or the events of a couple of days. If elephants in Uttaranchal are going to be properly counted and protected from the constant pressures of human ingress, then much more has to be invested in parks like Rajaji. The challenge of elephant conservation does not end with the creation of sanctuaries but requires sustained commitment on the part of state and national institutions, the scientific community, and donor agencies.

Raja and Yogi

In addition to the highways that run through Rajaji Park, there is a railway line connecting Hardwar and Dehradun. Between twenty and thirty trains travel this route each day, many of them passing through the park at night or early in the morning. After crossing a bridge on the outskirts of Hardwar, the railway tracks enter the forest and for much of the hour-long trip to Dehradun there is jungle on either side. During the day it is a scenic journey, with the Himalayas to the north and bridges over streams like the Song and Suswa, which flow into the Ganga. Despite its picturesque beauty this railway line presents a serious danger to wildlife. Since 1987 eighteen elephants have been struck by trains, particularly at night. In 1991 the Mussoorie Express collided with two female elephants and both of them were killed. One of the cows had a calf, less than three months old, which remained by the dead mother's side and was rescued and adopted by the forest department.

A. J. T. Johnsingh, who inspected the scene of the accident a few hours after it occurred, believes that the calf was standing on the tracks and the two female elephants were trying to protect him. While the baby elephant was pushed aside, the cows were unable to escape the advancing train because of its speed and the steep embankments. Johnsingh suggests that the elephants were drawn to the tracks by food scraps and litter that passengers and railway caterers discard from trains passing through the park.

Theories have also been put forward that the animals are blinded by the headlights of the trains and that the engines' whistles are pitched at a frequency elephants cannot hear. Negotiations between park authorities, railway officials, and wildlife organizations have led to proposals for using train whistles that are more easily audible to elephants. Rescheduling and reducing the speed of trains passing through the park have also been recommended, but enforcement is difficult and the dangers remain.

The young survivor of the train accident now lives at Chilla. He has been named Raja and at five years of age his tusks have grown out. Zahur Ahmed and the other mahouts have begun training him, though he is still too young to ride. Raja is four feet tall and most of the time he is kept chained, for his tusks are sharp and he doesn't know his own strength. When I mentioned

Raja's name, Zahur shook his head in consternation, saying he is a temperamental and headstrong elephant.

"Last year, during the monsoon, Raja got loose. He wandered off toward the Ganga and jumped into the river. The water was especially high that time of year and he was swept downstream to the bathing ghats near Har-ki-Pauri. Fortunately the sluice gates for the canal were closed; otherwise who knows where he might have ended up? The people at Hardwar were amazed to see an elephant come splashing ashore but when they saw the chain around his ankle, they realized he must belong to the forest department. Somebody telephoned from there and I had to rush down and bring him back. Thank God he wasn't injured. If Raja had been hurt or if he had drowned, they would have thrown me in jail for negligence."

Between my first visit to Chilla and my return in June, another young elephant, named Yogi, had been rescued from the wild. This calf was about four months old, and according to the mahouts he had become separated from his herd while they were raiding sugarcane fields on the eastern edge of the park. He was being looked after by one of the charrawallas, Rais Khan, and a young woman named Kadambari Mainkar, who works as a volunteer for the Wildlife Trust of India. She had been living at Chilla for a month, while helping to care for Yogi. Every two hours he was given a feed of Lactogen, a common brand of baby formula. The calf drank the Lactogen greedily, his trunk raised and swallowing in gulps, as Kadambari poured the milk into his mouth. Within five minutes he finished off three liters of the formula.

When I asked Kadambari if the Lactogen was similar to his mother's milk, she shook her head. "No, it doesn't compare but the formula is easier for him to digest than fresh cow or buffalo milk, which is too rich for elephants." Kadambari said that when she first arrived at Chilla, the calf was emaciated and its ribs were showing. Now he looked well fed and seemed to have no complaints about his diet.

After his meal, Yogi was given a bath in a blue plastic tub. Though it was much too small for him to get inside, the calf kneeled down beside the tub while Kadambari poured mugs of water over him. Having been separated from his mother only six weeks earlier, the way in which Yogi had adapted to his life in captivity was remarkable. Most of the time he did not need to be tethered or restrained, except at night when he was kept in a stall with wooden bars. He seemed to crave human company and followed Rais and Kadambari wherever they went. At this age he still hadn't grown tusks, and his face had a flattened, tapered shape that would gradually become more prominent and rounded. Among the mahouts there was a debate whether he would grow up to be a tusker or a makhna. On the top of his head and along his back he had stiff bristles that stood up like a bad haircut. Occasionally he would suck on the end of his trunk, just as a human baby sucks on its thumb. Though Yogi lay down to sleep when he was tired, he had an irrepressible amount of energy and playful curiosity. Following his bath he kicked his tub about the yard, chasing after it in a game.

"He's playing football," said Zahur Ahmed.

Yogi seemed to be adjusting well and showed signs of good health, but it will be quite awhile before his future can safely be predicted. Another male calf, rescued a few years back, lived at Chilla for a year but died suddenly after a brief illness. Though an elephant is fairly resilient, there are many diseases that can be contracted from other animals. At Chilla the forest complex and elephant stables have a barnyard atmosphere, with chickens, goats, and cows, as well as dogs and cats. Most of these animals seemed to take no notice of Yogi, except for a black cow that sometimes charged him and had to be shooed away.

Arundhati barely acknowledged Yogi's presence even though

her stable was next to his. Everyone at Chilla agreed that she has no maternal instincts and would harm the calf. Kadambari said that Yogi had tried to approach Arundhati several times but she made threatening gestures. This seemed uncharacteristic, for female elephants in a herd often take care of each other's young. The mahouts' explanation for Arundhati's indifference was that she herself had never given birth.

The day after I got to Chilla, Kadambari was leaving for Delhi. Though she had come there with no previous experience caring for elephant calves, Yogi's good health and survival seemed largely due to her efforts and commitment. Equally important, Kadambari had developed a good rapport with the mahouts, particularly Rais Khan.

"It must be difficult to leave," I said. "I'm sure Yogi will miss you."

"No." Kadambari shook her head adamantly. "I've made sure he hasn't become attached to me and Rais knows how to take care of him."

Her stay at Chilla has confirmed Kadambari's ambitions of pursuing a career in wildlife management but she feels she needs an M.Sc. degree.

"I studied political science in college but when I finished, I couldn't see myself sitting for the IAS [Indian Administrative Service] exam or becoming a lawyer. Now that I want to work with animals, it's difficult because I don't have a science degree."

Though Delhi is her home, returning there obviously didn't appeal to Kadambari.

"Too much traffic and noise," she said, glancing across at the toll barrier where vehicles were still lined up.

Given all of the pressures that exist in Rajaji Park, it is inevitable that wild elephants will need to be rescued from time to time. Ordinarily a female would never abandon a calf like Yogi; instead she would be more likely to behave like Raja's mother, who died trying to protect him from the train. Nevertheless, as human beings and elephants compete for habitat, there will be situations where calves get orphaned or lost.

Because baby elephants are endearing and photogenic, there is a natural human impulse to protect and nurture these creatures. However, the problem of elephant conservation is much greater than simply rearing a rescued calf. The true challenge is to provide the environmental safeguards that limit the need to take these animals from the wild. There is a danger that the larger issues of conservation will be overshadowed by the immediate appeal of a baby elephant. Though the dedication and selfless efforts of volunteers like Kadambari are admirable, she would be the first to admit that saving the life of an elephant calf like Yogi isn't really a solution. It may be a necessary and rewarding part of wildlife management, but greater efforts must be made to keep these elephants secure in the wild, allowing them as much freedom of movement as possible, while limiting confrontations with human beings.

The Edict at Kalsi

At the western end of the Dehradun valley, where the Yamuna River flows out of the Himalayas, stands a rock about ten feet tall. Though it has been covered by a domed pavilion since 1912 to protect it from the elements, the rock has an ageless, indestructible quality. This solid chunk of pale granite must have formed when shifting continents collided several million years ago to create the Himalayas. The rock itself is shaped like a mountain, tapering to a cragged summit, its face etched with traces of igneous scripts, over which the alphabets of erosion have left their mark. On a sloping plateau just above this rock lies the town of Kalsi, surrounded by mango orchards and a forest of sal trees. The first range of the Himalayas rises several thousand feet above, a tilted wall of ridges through which the

Yamuna cuts its winding course. Only a few kilometers away lies the western boundary of Rajaji National Park.

Inscribed on the rock at Kalsi is a proclamation issued by the Mauryan emperor Ashoka in 250 B.C., one of eight stone edicts that survive today. These inscriptions mark the considerable extent of Ashoka's dominions, covering all but the southernmost tip of the subcontinent. Carved in the fourteenth year of Ashoka's reign, after he had defeated the last of his enemies in the Battle of Kalinga, these edicts reflect the emperor's conversion to Buddhism. Deeply troubled by the senseless brutality of war, he chose to renounce violence and seek a path of peace. The eight sets of inscriptions, each of which is virtually identical, promote the Buddha's teachings—a philosophy in which all forms of life are revered and the moral ideals of charity and self-renunciation are placed above the material demands of this world.

A framed sign on the wall of the pavilion gives a summary translation: "The first edict prohibits the slaughter and sacrifice of animals. The second provides a system of medical aid for both men and animals and records the sowing of herbs, the digging of wells and the planting of trees, while the third enjoins . . . the promulgation of the great moral maxims of the Buddhist Creed, viz., honouring of parents and religious teachers, liberality, tolerance, and kindness to animals."

Preserved for more than two millennia, the message could not be clearer or more succinct. Human beings are held responsible for the protection and conservation of all living species, including their own. Through these edicts Ashoka was addressing his subjects in 250 B.C. but his words could just as easily be a proclamation for our time.

Many of the sculptures that survive from Ashoka's reign, such as the famous Lion Capital that now serves as a symbol of the Indian republic, display a refined artistry. Yet the natural shape of the boulder at Kalsi has not been refashioned and the inscription follows the contours of its surface, etched in the stone like the indentations of ancient fossils. Aside from the parallel lines of Brahmi script, the only image chiseled into the rock appears on its northern face, directly to the right of the inscriptions. If you were not looking for this carving, the faint outline would be easy to miss, but once it catches your eye, there is no mistaking the elephant. Standing about twelve inches high, a regal bull coils his trunk between his tusks, as if he were preparing to charge. There are no signs of captivity, no ropes or chains, no ornaments or decorations. He is a magnificent creature, as perfect as the stone statue from Gharapuri island, or the frieze of elephants at the beach temple in Mamallapuram. Running my hand over the surface of the rock, I could trace the outline of the elephant with my fingers, following the profile rendered two thousand years ago.

Beneath the tusker is one word: Gajatame—the superlative elephant, a symbol of Buddha and the divine image that his mother, Queen Maya, saw in her dream. Gajatame is the white elephant who descended into her womb, as well as an earlier incarnation of the Bodhisattva whose legends are told in the Jataka tales. Even if the illiterate pilgrims who walked past this stone in 250 B.C. were unable to read the edicts for themselves, they would have recognized this elephant and understood its meaning. Beyond the words themselves, the message that Ashoka conveyed to the farthest boundaries of his empire was embodied in the messenger.

VIII

gajasutra

City of Flowers

*T*welve hundred kilometers downriver from Kalsi and Hardwar
lay Ashoka's capital, the fabled metropolis of Pataliputra in the
kingdom of Magadha. Today the city of Patna stands on this site
and serves as the capital of Bihar, one of India's northern states,
which encompasses the territory of several ancient kingdoms in
the Gangetic plain. Strategically located near the confluence of
the Ganga, Sone, and Gandhak Rivers, Pataliputra was founded
by a king named Ajatashatru who built a fort there in 493 B.C.
Greek writers who visited India two centuries later, following
Alexander's brief foray into the subcontinent, transcribed the
name of the city as Palibothra and were particularly impressed
by the large number of war elephants kept by the Mauryan emperor Chandragupta, Ashoka's grandfather. Under the Mauryas,
Pataliputra developed into an important center of trade and
served as the seat of an empire that extended over most of India.
The historian R. C. Mazumdar describes it as

> The greatest city in India. It was about nine miles in length
> and a mile and a half in breadth. The wooden wall of the
> city, probably built of massive sal trees, had sixty-four
> gates, and was crowned with 570 towers. Surrounding the
> wall was a ditch, "six hundred feet in breadth, and thirty

cubits in depth." The royal palace within the city was one of the finest in the whole world, and its "gilded pillars, adorned with vines and silver birds" extorted the admiration of the Greeks.

Pataliputra means "son of the trumpet flower," and this name comes from the story of a young Brahmin who sat on the banks of the Ganga beneath a patali tree *(Bignonia suaveoleus)*. Having failed to find a suitable wife, he was overcome with distress and desire. Sensing his loneliness, the tree broke off one of its blossoming twigs and gave it to the Brahmin, saying this would become his wife. That night, as the young man continued his vigil beneath the tree, the trumpet flower turned into a beautiful woman. Later, after they were married and his wife had given birth to a son, the Brahmin built a house for her on the spot and named it Pataliputra.

The kingdom of Magadha, too, has floral associations, and it was sometimes referred to as Palasa, after the name of a tree commonly known as "flame of the forest" *(Butea frondosa)*. One of the most spectacular flowering trees in India, it blooms during spring with an abundance of bright red blossoms that cover its limbs in a conflagration of color. In Hindi this tree is still called palas and its leaves are a favorite food of elephants. The neighboring kingdom of Champa, to the east of Magadha, also takes its name from a flowering tree *(Michaelia champaca),* which elephants relish, eating both leaves and flowers, as well as its bark.

Megasthenes, a Greek ambassador to Pataliputra during the reign of Chandragupta, praised the gardens of the Mauryan capital. He also described many unusual and fantastic creatures in India, including gold-digging ants and men who slept inside their ears, but most of Megasthenes' admiration and curiosity was reserved for the elephant. One of his stories relates how elephants have a special appreciation for flowers.

The attendants even go in advance of their elephants and gather them flowers; for they are very fond of sweet perfumes, and they are accordingly taken out to the meadows, there to be trained under the influence of the sweetest fragrance. The animal selects the flowers according to their smell, and throws them as they are gathered into a basket which is held out by the trainer. This being filled, and harvest-work, so to speak, completed, he then bathes, and enjoys his bath with all the zest of a consummate voluptuary. On returning from bathing he is impatient to have his flowers, and if there is delay in bringing them he begins roaring, and will not taste a morsel of food till all the flowers he gathered are placed before him. This done, he takes the flowers out of the basket with his trunk and scatters them over the edge of his manger, and makes by this device their fine scent be, as it were, a relish to his food. He strews also a good quantity of them as litter over his stall, for he loves to have his sleep made sweet and pleasant.

Hiuen Tsiang, the famous Chinese pilgrim who visited Pataliputra in the seventh century A.D., related a somewhat similar tale in which he describes how herds of wild elephants gathered flowers in the forest and sprinkled these on a stupa containing relics of the Buddha. "Some of them brought on their tusks shrubs [leaves and branches], others with their trunks sprinkled water, some of them brought different flowers and all offered worship [as they stood] to the stupa." Hiuen Tsiang also recounted stories of "perfumed" tuskers as well as descriptions of fragrant mountains that were metaphorically compared to elephants.

In Buddhist iconography, elephants and lotus blossoms are often paired together, notably on the magnificent stone gateways at Sanchi, a Buddhist monument in central India. Numerous images of elephants carrying lotus blossoms in their trunks or surrounded by the open petals of a lotus are carved into the pillars at Sanchi. There is also a stone panel in which Queen Maya is shown preparing to give birth to the Buddha, while two elephants shower her with water. Each of these figures stands upon a lotus, offering a mirror image of the Hindu goddess Lakshmi, who also rises out of a lotus and is attended by elephants with lotus blossoms in their trunks.

The pleasure gardens of Pataliputra were irrigated by the Ganga and pools of water were filled with lotuses, ranging in color from pure white to livid pink. Other fragrant flowers, like jasmine and roses, were planted in beds. One of the ornamental trees often referred to in lyrical descriptions of the city is named after Ashoka *(Saraca asoca)*. A relatively small tree, just over twenty feet tall, it has sweet-smelling orange flowers that bloom in clusters, covering the dark crown of leaves in a shower of gold. Ashoka trees are considered sacred to Kama, the god of love, and according to legend they only bloom when kicked by a beautiful woman.

The Indian Tourism Development Corporation hotel where I stayed in Patna is called the Pataliputra Ashok, as if to suggest that the overpriced luxuries of today reflect the privileged grandeur of the past. Having arrived on a popular date for weddings, I had difficulty finding a room and little choice of accommodation. Most of the hotels in Patna were full of marriage parties, including the Chanakya and the Maurya, where uniformed bands stood ready to serenade the brides and grooms. Garlands of marigolds, strings of jasmine, and rose petals decorated the lobby of the Pataliputra Ashok, filling the air with a cloying fragrance. Later in the evening I watched a groom arrive in a blaze of blinking neon lights, accompanied by the raucous tunes of clarinets, trumpets, and trombones. Instead of riding on an elephant, as he might have done in ancient India, the bridegroom arrived in a Maruti van, festooned with marigolds and other flowers.

The Mauryan Empire (321–185 B.C.) represents one of the high points of Indian civilization, the culmination of a gradual process of conquest and settlement that began with the first Aryan invasions, around 1500 B.C. These migrant herdsmen from west and central Asia brought their cattle and horses across the north-western Himalayas and slowly moved eastward in search of new pastures. In many ways it could be argued that the spread of Indo-Aryan culture throughout the Gangetic plain was essentially a quest for fodder. This region is still referred to as "the cow belt" and cattle farming remains an important aspect of Bihar's rural economy. It is also no coincidence that the ex-chief minister of Bihar, Laloo Prasad Yadav, is embroiled in a scandal over the mis-appropriation of government subsidies for feed and fodder.

Following the course of the Ganga downstream, the Indo-Aryan herdsmen began to till the land and rapidly dominated the indigenous population, some of whom were farmers while others were hunters and gatherers. As an agrarian culture coalesced, scattered settlements grew into villages, towns, and cities. The Ganga allowed for transportation and commerce, as well as pro-viding a vital source of irrigation. Tribal republics and kingdoms emerged along with a stratified social structure of priests, war-riors, traders, and farmers, which developed into the caste system of today. This excluded most of the original inhabitants of the region, who were considered outcasts and untouchables by the Indo-Aryans.

At the same time, the process of deforestation had begun. While land was cleared for fields, the hardwood trees growing along the shores of the Ganga were felled to build the stockade walls of cities like Pataliputra. Except for a few stone structures, most of the buildings during this period were constructed out of wood, timber being the most readily available material. As the natural habitat of elephants was destroyed, the wild herds re-treated to the outer reaches of Magadha and beyond. In order to fill their elephant stables, the kings and emperors of Pataliputra were forced to acquire these animals from distant sources.

Within the Mauryan Empire eight gaja vanas, or "elephant forests," were identified and protected. These can be considered the first wildlife preserves in India. Only hunters licensed by the emperor to capture elephants were allowed to enter the forests. The most prized animals came from the Pracya Vana, east of Magadha, in sections of what is now West Bengal and Assam, and from the Kalinga Vana, to the south of Magadha in Orissa. If one looks at a map of these gaja vanas, they correspond to some of the wildlife sanctuaries of today. However, the territory that remains under forest cover is minuscule compared to the original elephant forests. Hardly 5 percent of the present state of Bihar is covered in forest, and that, too, is severely degraded.

The list of elephant forests appears in Kautilya's *Arthashastra,* a manual on statecraft written by an influential Brahmin in Chandragupta's court from 321 to 297 B.C. To serve the emperor and carry out his orders, Kautilya set up an elaborate bureaucracy, listing a number of overseers or superintendents responsible for the resources of the state. There was a superintendent of weights and measures, a superintendent of agriculture (responsible for the cultivation of flowers, as well as grains and vegetables), a superintendent of liquor, and a superintendent of prostitutes. Separate overseers were also appointed for cows, for horses, and for elephants. The last of these was known as the *hastiadhyaksha,* for whom Kautilya offered clear injunctions:

> The superintendent of elephants shall take proper steps to protect elephant forests and supervise the operations with regard to the standing or lying in stables of elephants, male, female, or young, when they are tied after training, and examine the proportional quantity of rations and grass, the extent of training given to them, their accoutrements and

ornaments, as well as the work of elephant doctors, of trainers of elephants in warlike feats and of grooms such as drivers, binders and others.

The precise dimensions of the elephant stables are noted by Kautilya, as well as instructions for their care and cleanliness. To provide adequate space, each elephant was housed in a stable twice as broad and twice as high as the animal. Stable floors were to be made of wooden planks, with appropriate drainage and provisions for the removal of dung. A daily schedule for elephants was prescribed: "The first and the seventh of the eight divisions of the day are the two bathing times of elephants, the time subsequent to those two periods is for their food; forenoon is the time for their exercise; afternoon is the time for sleep; one-third of the night is spent in taking wakeful rest."

Kautilya lists the specific rations issued to elephants, in proportion to their size. They were fed three different kinds of fodder—meadow grass, ordinary grass, and straw. This was supplemented with a variety of delicacies including rice, oil, butter, salt, curd, and milk, as well as quantities of sugar "in order to render the dish tasteful." In addition, the elephants were given broth, meat, and liquor, to prepare them for battle. The *Arthashastra* also indicates religious rituals—artis, pujas, and sacrifices—performed "for the safety of the elephants."

The capture of wild elephants was of particular importance to Kautilya and he is very specific about the age and temperament of animals caught in the wild. Twenty years was considered the ideal age for capturing an elephant and tuskers were clearly preferred over females. Most important of all was the recognition that a ready supply of elephants for the royal stables depended on the conservation of their natural habitat.

The superintendent of elephant forests with his retinue of forest guards shall not only maintain the upkeep of the forests, but also acquaint himself with all passages for entrance into, or exit from, such of them as are mountainous or boggy or contain rivers and lakes.

Whoever kills an elephant shall be put to death. The victory of kings [in battle] depends mainly upon elephants, for elephants being of large bodily frame are capable not only to destroy the arranged army of an enemy, his fortifications and encampments, but also to undertake works that are dangerous to life.

Historian Thomas Trautmann has argued that one of the principal reasons for Chandragupta's success in spreading the Mauryan Empire across northern India was his monopoly on elephants and horses, both of which were essential to victory in war. The armies of this period had four components: an elephant corps, cavalry, horse-drawn chariots, and infantry. According to Megasthenes, within Chandragupta's domain only the emperor was permitted to own an elephant. Though elephants were associated with royalty well before Chandragupta's time, the exclusive relationship between king and tusker seems to have been firmly established under Mauryan rule. Trautmann also reveals that elephants were exported by the Mauryas to the Greek rulers of Persia and Bactria in exchange for horses, one of the earliest examples of an international arms trade.

Having emerged from a cattle-driven culture, these descendants of itinerant herdsmen retained a close association with their livestock. Even as the land was cleared and plowed, Indo-Aryan culture continued to revere the cow and the horse, through rituals of worship and sacrifice. Not only were these animals an important economic resource, but also a symbol of conquest, best illustrated by the Asvamedha ritual in which a white stallion was set free to roam across the land. All of the territory over which the horse grazed was claimed by the king.

The taming of elephants, however, represents a different tradition. Whereas cows and horses were brought from west and central Asia, elephants belong to the forests of India and Southeast Asia (though evidence shows in earlier times they may have ranged as far as Mesopotamia, in present-day Iraq). The capture and training of elephants by Indo-Aryans suggest a shift from a migrant culture that brought its animals with it to a society that was now fully engaged in a new environment. Equally significant is the fact that the expertise required for capturing and training elephants was possessed by the indigenous people of this region, who were employed by the kings as elephant hunters and handlers. The *Rig Veda,* oldest of Sanskrit texts, reaching back to the earliest stages of Indo-Aryan culture, contains a reference to "hunters tracking wild elephants," who are compared to supplicants offering oblations at dawn and dusk. These forest-dwelling hunters provided an essential service, but their social status remained low and they were kept outside the caste system. Even today most mahouts and elephant handlers in India come from marginalized communities.

The fact that elephants became the sole prerogative of kings, as well as symbols of royalty and divinity, shows the reverence assigned to these beasts. Comparisons can be made with the buffalo, another indigenous animal that was domesticated. Whereas elephants were viewed with awe and veneration, buffaloes were generally linked to the darker side of nature and in mythology they often take the form of a demon that must be destroyed. The metaphorical and spiritual link between animals and power is a fundamental equation in almost every culture. In Hindu mythology, when the primordial ocean was churned by the combined forces of gods and demons, many wonderful objects and creatures emerged. Among these "jewels" of creation were three animals: Surabhi, the cow of plenty whose udders never run dry; Uchchaishravas, a pure white stallion, and Airavata, the elephant who carries Indra, Lord of the Universe, on his back.

Today India's most sacred animal may be the cow, but the struggle for power on the subcontinent from the fourth century B.C. until the eighteenth century A.D. depended largely on horses and elephants. The relationship between these two animals reveals a dichotomy of domination—while invaders arrived on horseback, they were only able to retain power after mounting an elephant. In other words, the conquest of India may have been achieved in the saddle but the seat of enduring power was the howdah. Whether it was the Indo-Aryans or the Mughals, *Equus* brought them across the Hindu Kush while *Elephas* permitted them to stay.

This dependence on animals is central to both mythology and history. By taming a wild creature, human beings assert their dominance over the natural world and at the same time accept responsibility for its care and maintenance. As the Ashokan edicts proclaim, it is deemed morally correct and commendable to treat animals with the same respect given to human life. Even though these ideals have not always been put into practice and are difficult to reconcile with the conditions under which many animals live today, the concept is clearly ingrained in the ethos of India.

Both Buddhism and Jainism began in the central Gangetic plain, most of which now lies within the modern state of Bihar. One of the fundamental tenets of these religions is a respect and reverence for all forms of life. This comes from a belief that every living creature is part of a collective divinity, a universal soul that inhabits a multitude of different forms. There is also a recognition that animals and human beings often exchange souls through transmigration and rebirth. Whether it is the Vedic hymns and epic narratives of Indo-Aryan cowherds, sermons of the Buddha, or Ashokan edicts, there is no question that Indian culture is grounded in a symbiotic relationship between human beings and other mammals.

The Sonepur Mela

"You should buy her."

"What am I going to do with an elephant?"

"Ride on her back."

"I've already got a car."

"Yes, but the fuel consumption is much better. With an elephant you won't spend money on petrol."

Like all good salesmen, Rajgiriji had his lines rehearsed and the small crowd gathered around us laughed at his joke. Vijaylakshmi, the elephant in question, was chewing on a mouthful of straw and didn't seem to share their amusement. She was an elderly female—I would have guessed her age as sixty, though Rajgiriji claimed she was thirty-five. I felt sure he was subtracting at least twenty years, but of all the elephants at the Sonepur cattle fair Vijaylakshmi had caught my eye. There was a mature beauty about her that the younger cows lacked and a self-possessed dignity. Her expressive face and ears still had traces of pink decorations, the smudged lines of lotus blossoms on her cheek, like an actress who hadn't removed all of her makeup.

"How much would she cost?"

Rajgiriji eyed me cautiously and smiled. He wasn't about to reveal a price, especially with so many others around.

"What would you offer?"

"I have no idea."

"She's very well behaved. Very gentle."

"I can see that. What about a price?"

"That's up to you."

Rajgiriji must have guessed that I wasn't going to buy his elephant, but he played along as there were no other customers. Earlier, he had told me that they had come from Madhya Pradesh,

about 400 kilometers away. Vijaylakshmi had walked all that distance and it had taken them over three weeks to reach the fair.
Rajgiriji claimed he wasn't sure if he wanted to sell his elephant,
saying, "I've only come to show her off." Most mahouts are employed by elephant owners, but he made it clear that Vijaylakshmi belonged to him. With a graying beard and long hair raked
back over his shoulders, Rajgiriji was dressed as a mendicant in
saffron robes. Traveling along the road to Sonepur, he and his
elephant had been able to live off the charity of others.

"When people see an elephant passing through a town or
village, they will often give her something to eat. But here at the
fair it is different. I have to buy her food, mostly sugarcane. It's
very expensive. You can see that she is hungry."

As if to prove his point, Vijaylakshmi stretched out her trunk
and tried to snatch the notebook from my hands, thinking it was
something she could eat. I took a step back but her trunk tugged
at my sleeve and sniffed expectantly.

Rajgiriji told me that they would be returning home in another day or two since he couldn't afford to stay at Sonepur
much longer. He complained that this year nobody was buying
or selling elephants at the fair because the government had made
it more expensive by raising fees and taxes.

"It costs almost 20,000 rupees just to process all the papers."

Most of the curious crowd had dispersed, and Rajgiriji invited me to sit with him and his companions near a fire they had
lit to brew some tea.

"I haven't decided if I want to sell her," he said again. "She
has been with me for so many years. I will take her back to Madhya Pradesh."

Tethered close by was a younger female who was much more
active than Vijaylakshmi, moving restlessly from side to side and
throwing straw onto her back so that she looked as if she were
covered with thatch.

Trying a different approach, I pointed to this elephant and

asked what sort of price she was likely to fetch. This time one of the other men replied.

"It all depends on what is agreed. Some elephants can go for two lakhs [200,000 rupees], others for four lakhs [400,000 rupees]. The big tuskers will cost more."

Buying and selling elephants is an uncertain business and those who engage in the trade are by nature taciturn and cautious. Ambiguities regarding an acceptable price seem to have remained constant from the time of the *Matangalila,* in which it is stated: "When one price is approved by both buyer and seller, that shall be known as the best price; what is disapproved by one of the parties, as a middling price; what is disapproved by both, as a bad price. Hence determining all by many careful experts, the price of elephants shall be arrived at." Though there can be no specific guidelines for purchasing an elephant, experienced elephant traders look at the age, the size, and demeanor as indications of the animal's value.

In his book *Travels on My Elephant,* the English writer Mark Shand describes his experiences at the Sonepur Mela, when he tried to sell his elephant Tara. Negotiations were protracted and evasive, while bids were made surreptitiously. According to Shand, one of the traditional methods of determining a price is for the buyer and seller to sit facing each other, their hands covered by a blanket. Without saying a word they press the knuckles on each other's fingers to indicate the amount they are willing to offer or accept. The advantage of this system is that nobody but the two men involved in the transaction knows what price has been agreed upon. As Shand discovered, however, when bargaining fails, strong-arm tactics are sometimes used. He was physically threatened by elephant traders who tried to intimidate him into selling Tara at a lower price. In the end he was forced to get police protection for his elephant and ultimately gave her away to a wildlife camp at Kanha National Park in Madhya Pradesh.

––––––––

The village of Sonepur lies twenty-five kilometers northwest of Patna. Most of the year it is a quiet, nondescript settlement on a wedge of land extending into the sandy floodplains at the confluence of the Ganga and Gandhak Rivers. During the Hindu month of Kartika, which usually falls between late November and early December, Sonepur undergoes a dramatic transformation, serving as the site for the largest cattle fair in India. Cows, oxen, bulls, buffaloes, horses, donkeys, mules, and camels are all brought to the Sonepur Mela. Here they are exhibited, haggled over, and traded in a commercial festival that recalls the traditions of Indo-Aryan herdsmen who migrated with their animals across the Gangetic plain over three millennia ago.

One of the chief attractions at the fair is the Hathi Bazaar, the only place in India where elephants are bought and sold in large numbers. As recently as two decades ago, over 300 elephants were brought to the Sonepur Mela and early in the twentieth century as many as 1,000 attended the Hathi Bazaar. This year there were only 77 elephants, and by the time I reached Sonepur, during the second week of the fair, many of these had already departed.

A mango orchard near a bridge across the Gandhak River is the designated exhibition ground for elephants. When I first arrived at nine o'clock in the morning, about thirty animals were tethered under the trees. Clusters of mahouts and charrawallas squatted together around cooking fires, smoke drifting up into the branches of the trees. From a distance the elephants looked like a placid herd, grazing amidst a protected grove, but as I approached, it became clear that each of them was chained to wooden pegs in the ground or tethered with ropes. Most of the elephants swayed from foot to foot, nibbling on bits of straw and sugarcane. Off to one side, about fifty meters from the other animals, was a handsome bull elephant with brass knobs fixed on the ends of his tusks. Though restrained by more than a dozen ropes, he stood defiantly erect and watched me with a belligerent expression. One of his attendants said the bull had recently gone into musth, though I could see no excretions from his temporal glands. The handler assured me that he was dangerous and a week earlier this tusker had broken loose, creating panic in the camp.

On the opposite side of the orchard stood another male who was also in musth. Dark streaks of fluid were visible on his cheeks and he was straining against his chains in anger and desperation. This elephant was at least a foot shorter than the other bull and had only one tusk, which gave him a mean, lopsided appearance. Both of his hind legs were chained together so that he could hardly move and his forelegs were also tied with ropes. When I circled around him to get a photograph, he strained to keep me in his line of vision and leaned forward as if trying to uproot the tree to which he was tied. His mahout was equally ill-tempered, a harassed-looking man, his turban unravelling and his clothes streaked with mud.

When I asked him how the elephant had lost his tusk, the mahout glared at me.

"How should I know? The *bhanchut* [sister fucker] must have rammed it into something. He's a *goonda hathi* [a rogue]."

I had been warned by several people that the elephants at Sonepur have a reputation for being violent. The implication is that they are brought to the fair from far away, because anybody who is familiar with their temperament would be unwilling to buy these animals. Whether this is true or not, there is always a danger that some of the elephants will be poorly handled and may go out of control. Just before my arrival, a tusker who was being brought to the fair went on a rampage, overturning a pickup truck and trampling a woman to death. Eventually, his mahout was able to regain control of the tusker and he returned home without attending the fair.

Tragedies like this are not uncommon at the Sonepur Mela, where handlers and spectators have been killed in the past. A large part of the problem is the huge crowds of people at the exhibition grounds. Such a concentration of animals and human beings, as well as the frenzied atmosphere of the fair, leads to inevitable accidents.

Dr. N. V. K. Ashraf, of the Wildlife Trust of India (WTI), has plenty of experience treating the illnesses and ailments of elephants, as he and his colleagues operate veterinary camps in Delhi and Jaipur. For the past two years Ashraf has attended the Sonepur Mela, providing free medical care to the elephants at the Hathi Bazaar. He feels that most elephants in captivity are not properly maintained. Often they are overworked and kept in an environment that is unsuitable, with inadequate water and little shade.

"The owners and mahouts depend on them for livelihood. They will make them walk and carry loads even when injured. Foot infections and sore backs are a common problem."

Loss of vision is another ailment elephants commonly suffer in captivity. Their eyes are particularly sensitive and prone to inflammation and infections. One of the young elephants brought to the Sonepur Mela this year had a gooey discharge streaming from his eyes that spattered in all directions every time he shook his head. Cataracts are prevalent in older animals. Of the ninety-two elephants that Ashraf examined at the fair in 2001, four were completely blind and an equal number had lost the sight of one eye. Twelve others suffered corneal opacity, the first symptom of blindness.

One of the challenges that Dr. Ashraf faces is the widespread suspicion among mahouts of modern medical techniques. Though he acknowledges that there are effective herbal remedies and treatments, handed down from one generation of elephant handlers to the next, Ashraf explains that antibiotics and other medicines can be much more beneficial. Through the WTI clinics, he and his colleagues disseminate information about the proper maintenance and care of tame elephants as well as their conservation in the wild. The data he has collected, along with records kept by the Bihar Forest Department, which monitors the Hathi Bazaar, provide a useful measure of the conditions under which elephants are kept in captivity.

Many of the elephants treated at Sonepur suffered from anemia, generally a result of poor diet. Though elephants are eclectic eaters, they require a nutritional balance not always provided by their owners. At the Sonepur Mela, for instance, most of the animals were fed sugarcane and occasionally whole wheat bread. Little or no green fodder was available, though it is essential for their well-being.

Ashraf also pointed out that worms and other intestinal parasites weaken and debilitate elephants. With a digestive tract that processes over 200 kilos of food a day, they are particularly vulnerable. Very few of the owners deworm their animals regularly or give them medicines to treat liver flukes, a chronic illness that often results from elephants living in close proximity with other livestock. One of the sure signs that an elephant has worms is when it eats dirt, which serves as a natural purgative but does not cure the problem.

Injuries sometimes occur when elephants are being transported to the Sonepur Mela. Rather than walking their animals over long distances, many of the owners load them onto trucks and have them driven to the fair. If the elephants do not have adequate space and are confined in the trucks for too long a period, their health can be seriously affected. Ashraf described how one elephant suffered from a dislocated tusk from riding in a truck. The tusk was cracked and hanging loosely from its socket, a sure sign that it would eventually fall out.

At the main entrance to the Sonepur fairgrounds stands a statue of an elephant whose leg is caught in the jaws of a crocodile. The tusker is rearing up with its trunk in the air, trying to escape, while the crocodile twists its tail and body in a ferocious struggle. Rising above these two creatures is an image of the god Vishnu, surrounded by a fiery halo. In his role as protector and sustainer of the universe, Vishnu keeps the world in balance and intervenes in the never-ending conflict between good and evil.

The statue at Sonepur is a recent construction made of cement and plaster, but it illustrates an ancient myth from the *Bhagavad Purana*. An elephant king, known as Gajendra, had recently gone into musth and was sporting with his many wives near a lake covered with lotus blossoms. In a playful mood, the elephant king plunged into the water to cool himself and quench his thirst. An enormous crocodile inhabited this lake and before he knew it, Gajendra felt his leg clamped in a vicious grip. Both animals were equally matched in strength, and they fought with each other in a desperate battle that churned the waters of the lake and lasted over a thousand years. Finally, as the tusker's strength began to wane, he raised his trunk and prayed to Vishnu for help. Immediately, the benevolent god appeared and unleashed his most powerful weapon, the Sudarshan Chakra, cutting off the crocodile's head.

On one level this myth is a simple allegory about the power of prayer and devotion, but as with most stories in the *Bhagavad Purana,* there is another layer to the narrative. After Vishnu rescues Gajendra, it is revealed that this tusker was actually a king named Indradyumna, who had been changed into an elephant by the curse of a sage. The crocodile had been a gandharva, or "demigod," named Huhu. He had also been the victim of a sage's curse. Many years earlier, Huhu had been frolicking in the lake with his beautiful consorts, when Rishi Davala arrived to take a bath. Caught up in his revelry, Huhu mischievously grabbed the sage's legs and pulled him underwater. Davala was so incensed that he angrily transformed Huhu into a crocodile.

The historical associations between the Gajendra Moksha myth and the elephant market at the Sonepur Mela are not entirely clear. The fair is more closely linked to bathing rituals performed on Kartika Purnima, the full moon with which the month commences. On the first few days of the festival, crowds of pilgrims and elephants bathe near the confluence of the Gandhak and Ganga Rivers, while prayers are offered at the Harihar Temple, dedicated to both Vishnu and Shiva. Though there may well have been crocodiles at Sonepur in the past, they no longer inhabit these rivers, and today both the pilgrims and elephants are safe.

By midday the Sonepur Mela had gathered momentum. Leaving the Hathi Bazaar, I wandered about other sections of the fair. Each species of livestock has its own exhibition ground, areas of which are covered with open tents that serve as temporary stables. Cows were the most plentiful animal, and there were separate enclosures for various breeds, from the ubiquitous white cattle that can be found in every Indian village to more exotic Jerseys and Herefords. Oxen and buffaloes also occupied their own territory at the fairground, with segregated stalls and pens. The horses were kept in a mango orchard about half a kilometer away from the elephants. Here the scene was similar to the Hathi Bazaar, as the animals stood in lines, tethered to wooden pegs, while their owners lounged about on string cots discussing the merits of each beast. White stallions held pride of place and as I watched one being brushed by his groom, the owner offered to let me take him for a ride. I quickly declined, for the horse appeared so high-strung that I wouldn't have been able to stay on his back for more than a few seconds. His owner told me that they had come all the way from Agra and like the elephant handlers he, too, was disappointed in sales this year. He suggested that it may have had something to do with the dates for the fair coinciding with Ramazan, the Islamic month of fasting. In northern India many owners of elephants and horses are Muslims, who may have chosen to stay away this year. Though he himself would not eat or drink until after sunset, the horse trader insisted on ordering a cup of tea for me from a stall nearby. His horses, too, were well supplied with food and most of them had feed sacks over their heads.

In addition to livestock there is a section of the Sonepur Mela where pets of all kinds are sold, including dogs and cats of dubious pedigree. The miserable conditions under which these animals are kept almost make it seem as if the elephants and horses are pampered. Squeezed into crowded pens with limited water and food, they seem cursed to endure the abuses of mankind. Cages full of birds—budgerigars, finches, pigeons, and partridges—attract those who fancy these pets. At the entrance to the market sat a fortune teller with a trained parakeet, who picked up tarot cards with its beak. For ten rupees this bird, whose wings were clipped, judiciously circled the cards, then selected one and flipped it over to reveal secrets of the future.

The fairgrounds at Sonepur extend in all directions, and there is a chaotic, jumbled atmosphere, despite efforts by state authorities to organize and regulate the fair. A number of stalls promoting agricultural innovations, from new seeds and fertilizers to improved feed and fodder, were sponsored by the Bihar government. These seemed to draw the thinnest crowds as most people gravitated toward hawkers selling sweets and snacks—heaps of knotted jalebis dripping with syrup and plates of potato chaat, with green chilis sticking out like horns. Tented arcades were full of stalls selling plastic toys and costume jewelry. Rows of glass bangles were on display, enough to ornament the wrists of thousands of yakshis.

Three large Ferris wheels spun overhead and there were many other rides, including one with brightly painted helicopters that looked suicidal. Dizzy children staggered off these rides, while others eagerly lined up for the vertiginous pleasures of being hurled about in machines designed to shake loose every bone in the human body. For the aspiring marksman, there were plenty of shooting galleries with balloons as targets and ring toss booths where the prizes ranged from cheap plastic dolls to more useful things like Eveready flashlights.

A tractor pulling a tank of water made its way through the fairground with a sprinkler system to lay the dust. The subtle scent of wet earth mingled with the gingery tang of spices being sold in gunnysacks and the heady perfumes of hair oil and fragrant essences displayed on a hawker's cart. All of these smells combined with the pungent odor of onions frying in mustard oil at the food stalls. Throughout the afternoon crowds swarmed through the lines of booths until at dusk the festivities seemed to reach a climax. Just before sunset, banks of colored lights flickered on and the loudspeakers, which had been playing film songs throughout the day, seemed to increase in volume.

The Hathi Bazaar, however, was silent. In the gray twilight, two elephants came up from the river after taking their evening bath and returned to their places in the mango orchard. The crowds of curious spectators had departed, and before settling down for the night the handlers checked chains and ropes, then added an armload of sugarcane to the heaps of straw. One or two electric bulbs had been rigged up in the branches of the mango trees, but most of the light came from smoky cooking fires, where the mahouts sat together, wrapped in shawls and blankets. Voices were muffled and once in awhile there was laughter. In the flickering aura of firelight several of the elephants swayed back and forth, but most of them were subdued, ignoring the clamor and flashing lights of the fairground.

Daughters of the Trumpet Flower

"Indian women, if possessed of uncommon discretion, would not stray from virtue for any reward short of an elephant, but on receiving this a lady lets the giver enjoy her person. Nor do the Indians consider it any disgrace to a woman to grant her favour for an elephant, but it is regarded as a high compliment to the sex that their charms should be deemed worth an elephant."

These are the observations and opinions of Nearchus, one of Alexander's military commanders. Even in the fourth century B.C., surrendering one's virtue for an elephant must have been a rare transaction—more Grecian fantasy than Indian custom. Nearchus's statement, however, may explain the enthusiasm with which Greek satraps, who governed kingdoms along the western margins of the subcontinent, struck coins bearing the images of tuskers. In the city of Pataliputra seducing a woman with the gift of an elephant would have been severely restricted by an imperial monopoly.

As an emblem of desire, the image of gajagamini—a woman whose walk is as seductive as an elephant's—can be found in many ancient texts, including the *Ramayana*. When the demon Ravana first sees Sita, the wife of Rama, he is overcome with lust and asks, "Who art thou, golden woman, clad in yellow silks, wearing like a lotus pond the bright garland of lotuses? Are you Modesty, Beauty, Fame, the Good Goddess of Luck, a celestial Nymph, oh bright-faced woman? Or Prosperity, my large-buttocked one, or Lust that freely rambles? . . . Your buttocks are broad and firm, your thighs tapering like the trunk of an elephant."

Offended by Ravana's lechery, the chaste Sita rebuffed his advances. On the other hand, courtesans in Pataliputra probably would have taken these words as a compliment and recognized in them the romantic conventions of narrative verse.

Without question, the most widely translated Sanskrit text in the world is the *Kamasutra*, possibly composed in Pataliputra during the Gupta Empire, around the third century A.D. As a catalogue of sexual postures and erotic etiquette, it is a remarkable book, though Vatsyayana, the author, also provides an invaluable perspective on society and culture in ancient India. While sex is one of the few pleasures that crosses the barriers of class and caste, the refined art and science of love prescribed in the *Kamasutra* would have been practiced and enjoyed only by a leisured elite.

The courtesans of Pataliputra were sometimes referred to as Pataliputrika—"daughters of the trumpet flower." Different classes of women were trained in seduction, from ordinary prostitutes to the most sought-after and all-but-inaccessible beauties of "uncommon discretion," sometimes referred to as "courtesans de luxe." As in the *Arthashastra* there is a penchant for categorization in the *Kamasutra*. Like Kautilya's instructions for the games of politics and power, Vatsyayana's rules for making love are equally precise and codified, as are the physical attributes of each amorous couple.

Varying dimensions of human genitalia are classified according to corresponding members of animal species. A man is ranked according to the length and circumference of his penis as either a hare, a bull, or a stallion. A woman's vagina is similarly compared to that of a doe, a mare, or an elephant cow. Vatsyayana then goes on to describe which pairs of sexual organs are best suited for intercourse. With the exception of the stallion and the mare—surprisingly, not equally paired—the other animals represent an assortment of different species. Perhaps most intriguing of all is the fact that male elephants do not figure in this genital hierarchy. One possible explanation might be that only a king or emperor could assert his royal prerogative by comparing himself to a tusker. The omission of bull elephants in the *Kamasutra* is all the more puzzling because Vatsyayana does compare certain women to the cow elephant who can only achieve an "equal coupling" with a stallion.

For a modern, urban reader these comparisons may seem crude and inexact, but within the context of early Indo-Aryan culture, where the relationship between men and animals was closely intertwined, these anatomical analogies seem perfectly appropriate. Rather than being a work of pornography or bestiality, as it is sometimes perceived, the *Kamasutra* simply compiles the various possibilities of sexual experience that can be learned from other species. As Vatsyayana explains:

A man who understands the heart should
enlarge his repertory of techniques for sexual ecstasy
by this means and that, imitating the amorous movements
of tame animals, wild animals, and birds.

In Indian mythology and literature, the consummate image of desire is the yakshi, a celestial nymph who epitomizes the ideals of female beauty. She is gajagamini—"seductive as an elephant." Yakshis lead men astray, like Ruchira, the mother of Palakapya in the *Matangalila,* who was turned into an elephant because her beauty distracted a meditating sage. Or like the anonymous yakshi in Kalidasa's poem *Meghaduta,* whose irresistible charms made her lover forget to close the garden gate before the elephant entered and trampled the flowers. The same yakshi receives passionate messages from her beloved through an elephant in the clouds.

Standing on a pedestal, her face is calm, almost impassive, with arched eyebrows and pursed lips, betraying no emotion. Pendant earrings frame her cheeks and the oval outline of her chin. Her hair is drawn back in an elaborate coiffure, adorned with strings of flowers and jewels. Bracelets cover her arms from wrist to elbow. Three strands of beads encircle her throat and fall between her breasts, pinched together like the confluence of rivers. Heavy and rounded, her breasts are swollen, the areolas and nipples delicately etched. Her waist is narrow but her hips are full, and below the dimpled navel lie three folds of flesh, like the pleats of her skirt. An ornate girdle around her hips emphasizes ample thighs, which can only be compared to an elephant's trunk as they taper to her feet. Beneath the hem of a cascading garment, her anklets look like heavy chains. Over one shoulder she carries a yak tail fly whisk, held casually in her right hand, as if preparing to tease the sultry air. The voluptuous contours of her body and the smooth lustre of her skin seem to glow in the artificial light.

Known as either *The Chauri Bearer* or *The Didarganj Yakshi,* this statue dates back to Pataliputra in the first century B.C. Carved out of pale amber sandstone with a highly polished surface, the yakshi is not so much a lifelike image of a woman but more a figure out of an erotic fantasy. She is known as *The Chauri Bearer* because of the fly whisk that she carries. Her other name comes from the place where she was found. Didarganj is a section of Patna, along the banks of the Ganga, where the statue was discovered buried in the sand.

Little is known about the history of this yakshi or where she may have originally stood, whether in a temple or a palace. With complacent anonymity, *The Chauri Bearer* has survived the ravages of time. A section of her hair has broken off and her nose is slightly chipped, but she retains her ancient charms. Sharing the gallery with her at the Patna Museum are a number of other statues, including the nude male torso of a Jain Tirthankara, carved from the same polished stone. This statue has been badly damaged and only the chest, belly, and genitals remain. Rather than being the yakshi's lover, however, he is a celibate saint. There is a second woman in the room, a carving from the third century A.D., when the *Kamasutra* was composed. She stands in a coy pose, looking down at a parakeet that sits by her feet, one of the companions of Kama, the god of love and passion. In another gallery of the museum are several terra-cotta figurines, labeled as THE EARTH MOTHER. With full breasts and broad hips, all of these images share a common profile of fecundity.

The Patna Museum is a warehouse of culture. Its color-washed exterior of yellow and red is typical of the Indo-Saracenic style favored by colonial architects—an ornamental facade suggesting exotic objects within. Though *The Didarganj Yakshi* is brightly illuminated, most of the galleries are poorly lit and the exhibits are badly labeled, so that visitors spend a great deal of time squinting through dusty glass cabinets and puzzling over cryptic captions.

The museum also has a natural history wing. Though it contains little evidence of elephants, other than a few pieces of ivory, there is a 200-million-year-old petrified tree. Its trunk is fifty-three feet long and stretches the length of a hallway. The preserved remains of this verdant giant are a reminder of long-lost forests that once covered the Gangetic plain. In the center of the gallery is a fully mounted gaur, as well as trophies of tigers, leopards, and bears. Moldering specimens of waterbirds, including a rare pink-headed duck, still bear the ravages of bird shot. The most unusual exhibit, however, is a stuffed specimen of two goat kids born as Siamese twins. Their bodies are joined in such a way that each of the eight legs and tiny hooves point in opposite directions. This exhibit is simply labeled A FREAK OF NATURE.

Marx at the Zoo

The Communist Party of India (ML) was holding its fourth All-India Congress in Patna and crossroads throughout the city were draped in red banners emblazoned with the hammer and sickle. As the bicycle rickshaw in which I was riding weaved through traffic, I could see groups of delegates walking along the roadside carrying party flags. Almost every wall was plastered with posters that bore slogans like RED REVOLT. In India there are several factions of the party, and the initials ML stand for "Marxist-Leninist," which is one of the more extreme brands of Communism, linked to the Naxalites, Maoists, and the People's War Group.

I was on my way to the zoo, which I felt sure would be a disappointment after the clamorous intensity of the Sonepur Mela. To my mind, visiting a zoo has always been an artificial, contrived experience compared to seeing animals in the wild. The atmosphere of incarceration, crowded cells full of isolated species or walled enclosures surrounded by stagnant moats, has never appealed to me. Even in the best zoos there is a feeling of confinement without the possibility of escape. Earlier, I had visited zoos in Delhi, Mumbai, and Mysore, all of which had left me feeling depressed.

The official name for the Patna Zoo is the Sanjay Gandhi Biological Gardens. Five rupees is the price of entry and an additional twenty rupees for a camera. Once inside, the first thing that struck me was the number of trees. After the concrete skyline of the city, I felt as if I were entering a forest or protected grove. Footpaths were neatly laid out, bordered by flower beds and ornamental trees.

A cage full of langur monkeys swinging on old car tires brought back my sense of ambivalence, as did a squalid aviary housing parakeets and budgerigars. A lion was roaring, its deep bass moan like a resonant complaint, though the enclosures in which these cats were kept seemed clean and well maintained. The leopards looked healthy enough and only one of them was pacing back and forth, its pendulous tail keeping time like a furry metronome. On ahead two tigers lay in the shade, their coats far brighter than the yellow fur on the moth-eaten specimens at the museum.

It was still early in the day, though a steady stream of people were passing through the turnstiles and wandering about the scattered cages. Most of the visitors didn't look as if they were residents of Patna, but were farmers from outlying villages or visitors from other towns. Many of the Marxist-Leninists had taken a break from their rallies and chosen to visit the zoo. Several of them greeted me with a friendly wave of their red flags. One of the groups told me they had come from West Bengal, another from Madhya Pradesh. When I spoke with them in Hindi, they immediately asked if I was Russian. The fact that I was an American led to some consternation and amusement, but the party activists seemed less interested in politics than enjoying the verdant surroundings of the gardens. Throwing arms around each other's shoulders or holding a comrade's hand, they strolled along a leafy promenade encircling a lake at the center of the zoo.

Watching the delegates, I had to question my own preconception of the zoo and its egalitarian pleasures. For five rupees I had seen more animals in an hour than I had observed in all my travels to national parks and wildlife sanctuaries, where the entry fees were a hundred times as much. The green space provided by the zoo in the center of Patna seemed to offer at least some semblance of a forest. And for the visitors, most of whom could never afford the cost of visiting a wildlife sanctuary, there was an opportunity to view these endangered animals.

Some of the strongest support for the CPI (ML) comes from areas of Bihar, Jharkhand, and Orissa, where a large tribal population lives on or near forest land. These Adivasis have traditionally relied on forest resources for their survival but are now confronted with an economy and social order that denies them access to this way of life. Invariably, the conservation of nature becomes a flash point in a class struggle between marginalized communities and state agencies like the forest department.

In a book titled *Ecology and Equity: The Use and Abuse of Nature in Contemporary India,* Madhav Gadgil and Ramachandra Guha have explored the social and political dimensions of various environmental issues, from mining and dams to timber harvesting and wildlife management. Among many different strains of environmental activism, the authors identify a group they call "ecological Marxists," who "see the problem in political and economic terms, arguing that it is unequal access to resources, rather than a question of values, which better explains the patterns and process of environmental degradation in India. In this sharply stratified society, the rich destroy nature in the pursuit of profit, while the poor do so simply to survive. . . . For ecological Marxists, therefore, the creation of an economically just society is a logical precondition of social and ecological harmony."

Gadgil and Guha go on to describe as "omnivores" the wealthy urban elite, who consume an unequal share of natural resources within an industrial-capitalist economy. At the same time these omnivores advocate the creation and preservation of national parks and sanctuaries. Whether it be for recreation, scientific research, or because of environmental and moral values, these forests are protected from the incursions of hunters, gatherers, and graziers. Denying these people an equal share of resources leads to many of the tensions associated with wildlife sanctuaries, such as poaching and overgrazing. Regardless of ideology, there is a great deal of truth in this analysis, and most conservationists in India now accept the idea that forests can only be preserved with the active involvement of forest dwellers.

The Patna Zoo may not be everyone's ideal vision of nature, but it does, in a very limited sense, offer a more equitable and accessible experience of India's wildlife than a national park. Peripheral sections of the zoo remain undeveloped and have been left relatively wild. Seeing a sign with an arrow pointing to a jungle trail, I followed it through a shisham grove to the elephant enclosure. In front of the entrance was a large banyan tree beneath which stood a temple. The shrine was tended by a stooped pundit with an unruly white beard. He explained that the temple was dedicated to Pashupati, which means guardian of animals, one of the many names of Shiva. The inner sanctuary of the temple was a masonry shrine, while the exterior was constructed of bamboo like a huge birdcage. It was decorated with ribbons of red cloth and silver tinsel that looked a bit like Communist Party banners.

Only one elephant lives at the Patna Zoo, a lone female named Mala, who was standing in her stall, feeding on straw. Though none of the caretakers was around, the gate of her enclosure was open and the priest led me inside. He seemed well acquainted with Mala and went across to a hand pump nearby and filled a clay vessel with water. Holding her trunk up like a funnel, Mala let him pour the water into her nostrils, then sprayed it into her mouth. The pundit also gave her some prasad from the temple, white sugary sweets that she obviously enjoyed.

Directly across from the elephant enclosure was a high fence that separated the zoo from the Patna Golf Club. The two areas were contiguous and the branches of trees extended from one side of the fence to the other. Carefully tended greens and broad fairways lined with trees presented a cultivated expanse of grass and foliage. The golfers and caddies were absorbed in their game and paid no attention to the animals and visitors in the zoo. A group of Marxist-Leninists, however, seemed intrigued by this unusual pursuit, pointing at the white balls that rolled across the close-cropped turf. Sauntering over to the fence, they stood and watched the golfers putting on the green as if they were a rare, omnivorous species, isolated within their own enclosure. After the foursome sank their putts and then teed off, they headed down the fairway to the next hole, oblivious of the eyes that followed them.

IX

power and pomp

Hathiwallas

ITO Bridge in New Delhi gets its name from the Central Government Income Tax Office that stands nearby, a bastion of bureaucracy overlooking one of the main arteries of the city. A steady cavalcade of traffic funnels in from the eastern suburbs, across the Yamuna River. Over a million commuters take this route every day, riding buses, cars, scooters, autorickshaws, or bicycles, their wheels spinning in unison and churning up clouds of exhaust-laden air. The constant rumble of engines, the shrill yelp of brakes, and the relentless carping of horns makes it seem as if the city is being assaulted by a motorized army of invaders who pour across the bridge each morning.

In June, with temperatures above 100 degrees Fahrenheit, heat radiates off the stream of vehicles like a warped mirage. From the far side of the bridge two gray shapes approach, blurred by the shimmering heat and diesel fumes. Almost as tall as the Delhi Transport Corporation buses that roar past them, these figures slowly advance, wading through the traffic. Only when they are halfway across the bridge can I clearly see the flapping ears and trunks swinging from side to side, as if to wave past impatient drivers. Amidst the frenzied traffic, these elephants seem oblivious of the mopeds and scooters that dodge around their legs. They have a different momentum from the rest of the city, as if out of step with the present age. Seated on their backs, the mahouts, too, look like figures from another era. Unlike the helmeted and goggled motorcyclists who careen from lane to lane, the riders on their elephants advance across the Yamuna at a stately pace.

Reaching the end of ITO Bridge, the two animals turn left out of traffic and trudge down a short dirt road to the riverbank. While cars and trucks continue to race by overhead, the elephants make their way to a rusty hand pump where the mahouts slide off their backs. A green plastic bucket is filled and refilled, though it hardly seems enough for the elephants to slake their thirst as they drain it with their trunks and squirt water down their throats. Half a dozen other elephants are standing nearby, some in the river, others feeding on straw.

The hathiwallas (elephant keepers), as they advertise themselves on business cards, have established a makeshift colony near ITO Bridge. Altogether, about thirty elephants are stabled here, though the actual number is difficult to pin down and could be higher. The elephants, decorated with costume jewelry and velvet trappings, are hired out for wedding receptions and other ceremonies. On occasion, they also carry tourists for joyrides. Gayyur Ali, one of the hathiwallas, was a short, rotund man who had rolled his T-shirt up over his stomach in an effort to cool off. Standing together in the shade of the bridge, we watched three elephants bathing in the Yamuna. A tusker and a makhna stood beside the pylons of the bridge, spraying themselves with water, while out in the middle of the river a female was swimming on her own. There was hardly any current but the water was over her head, and she disappeared beneath the surface several times, using her trunk as a snorkel. The elephants seemed to be enjoying their bath though the water was as black as engine oil and smelled like a cesspool.

"All of this is sewage from the drains of Delhi, which empty into the Yamuna," said Gayyur Ali in disgust. "Last year the government supposedly spent twenty lakh [2,000,000 rupees] to clean up the river. As you can see, it hasn't made any difference."

In addition to household effluents the Yamuna absorbs chemical waste from numerous factories, making it one of the most poisonous rivers in India, even when the pollution is flushed downstream by monsoon floods. That same morning, the newspapers had carried a story about thousands of fish dying in the Yamuna near Agra, their rotting remains washing up on the shore.

Despite the stench of the water and its oily hue, a group of young boys were also swimming near the bridge, jumping and diving into the river. Like the elephants they had no other place to cool off on a summer day. With very little shade and no clean water, the elephants at ITO Bridge share the miseries of Delhi's poorest residents. Their colony is one of many urban slums built along the margins of the city, where migrants to the capital have settled in hope of scraping together a better life.

Gayyur Ali complained that it was increasingly difficult for the hathiwallas to get fodder for their elephants. He pointed across the river to an area of grassland that floods during the monsoon. "Early in the morning, before dawn, we take our elephants over there to collect grass," he said. Far off in the distance I could see an elephant standing on the riverbank, where it had been taken to graze. On either side of the Yamuna rose the rooftops of the city, as well as the chimneys of the Indraprastha Power Plant that generates electricity for the capital.

"By four o'clock in the afternoon all of the elephants will be back here," said Gayyur Ali. "Most of our work is in the evening, attending wedding receptions."

The presence of elephants at a wedding has always been considered auspicious and they now serve as status symbols for the nouveau riche. While years ago Delhi's elephants attended upon kings and emperors, it is now the city's wealthy middle class that patronize these creatures. At most weddings the elephants simply stand at the entrance to a bride's house, welcoming the guests with ostentatious gestures.

"Tuskers provide the most imposing presence, but many people in Delhi are afraid of them and mostly they ask for females because they worry that males will injure their guests," said Gayyur Ali, patting the tusker bathing next to us to prove how tame he was.

"Of course, when he goes into musth, then we have to be careful and keep him chained." Brushing his hand against the tusker's forehead, Gayyur Ali showed me the tiny hole from which the temporal excretions flow. It was difficult to see, less than half a centimeter in diameter, like a pore in his skin.

"During musth we collect some of the fluid, if the elephant will let us. People pay a lot for musth because they believe it is an aphrodisiac." Cocking his head and grinning, Gayyur Ali continued. "Put a few drops in a glass of milk, then drink it up and you'll be full of energy all night."

Fresh from the polluted waters of the Yamuna and decorated with floral designs, these elephants proceed from the dusty slum in which they live to prosperous colonies like Greater Kailash or Panchsheel Park. Nothing less than living symbols of fertility and a reminder of Ganesha, they shower the happy couples with blessings.

In recent years, one of the most famous of Delhi's elephants was known as Lal Trikon, "the red triangle." He was employed in a family planning campaign in the late 1960s and early 1970s. During this period, the inverted red triangle became a nationwide emblem for population control in India and was accompanied by a picture of two parents and two children, with the slogan TWO OR THREE CHILDREN, THAT'S ALL! This symbol was conceived by T. K. Tyagi, media commissioner for the Department of Family Planning, who wanted to spread the message of population control throughout the country. Funded in part by the Ford Foundation, Tyagi hired one of the hathiwallas of Delhi to parade his elephant through the streets of the capital, draped with banners decorated with the red triangle. Those who rode on Lal Trikon's back distributed flyers and free condoms throughout the city. The success of this mass media campaign has been questioned along with other strategies of "social marketing," but there can be no doubt about the message that Lal Trikon conveyed. As a symbol of fertility he also promoted contraception.

Another hathiwalla I spoke with at ITO Bridge was Rashid Ahmed Khan. He told me that he used to be the head mahout at the Ashoka Hotel, one of Delhi's oldest and largest five-star hotels, where Khan's elephants carried tourists on their backs.

"Whenever VIPs came to Delhi, I would take my elephants to the airport and we would greet them. All of the world's leaders have had their photographs taken with me and my elephants."

A brusque, dignified man in his late fifties, with a twirled moustache, Khan looks very much like the Air India maharajah. Though the Ashoka Hotel no longer keeps elephants on retainer, Khan still hires them out to tour operators who want to give foreign guests a joyride. He also rents them for weddings, where he can charge as much as 15,000 rupees a night.

Khan said that most of the hathiwallas originally come from Gaziabad, a town across the Yamuna and about thirty kilometers from Delhi. For many years, they used to keep their elephants near the Red Fort in Old Delhi, beside the dhobi ghat where washermen still do laundry along the banks of the Yamuna. About thirty years ago, the hathiwallas shifted to ITO Bridge because it is closer to the colonies in New Delhi where most of their business lies.

"Things have changed for the worse," he said, with a note of regret and disgust. "The traffic in Delhi has become impossible and the pollution is worse than it ever was. We also have a lot of accidents. Just four days ago a truck collided with one of our elephants. It was only injured slightly but the boy who was riding on its back fell off and died. Three months earlier there was another accident. Both the elephant and the mahout were killed. It usually happens at night or early in the morning when they are returning from a wedding. As you know, drivers in Delhi are reckless and they don't expect to find an elephant on the road."

War and Conquest

As each successive wave of invaders crossed over the snow-capped ranges of the Hindu Kush and set forth to conquer India, they were met by another, equally forbidding obstacle. Around A.D. 1031 the Ghaznavid poet Farrukhi wrote: "One may ask, 'What are those 1,700 odd mountains?' I reply, 'They are the 1,700 odd elephants of the Shah.'" Thirteen centuries earlier, in a battle that has assumed mythic proportions, Alexander of Macedon confronted the war elephants of King Porus. Though the Greek army prevailed, Alexander's men were unwilling to carry on with his conquests, in part because of their fear of elephants. When weaponry consisted of swords, javelins, and arrows, an armored phalanx of tuskers bearing archers on their backs must have presented an image of invincibility. Hindu mythology contains a variety of stories that assert the elephant's role in warfare, beginning with the primeval struggle between gods and demons, in which the great tusker Airavata and his companions led Indra's forces into battle. During this conflict, which lasted thousands of years, the elephants began to show signs of panic and fatigue, at which point Brahma instilled them with the spirit of musth. Enraged and fearless, the elephants went on a cosmic rampage, carrying the gods to victory.

War elephants are also mentioned in the *Mahabharata* epic as the Pandava and Kaurava cousins fought for control of their ancestral domain. The capital of this kingdom was Hastinapur—literally, "the elephant city"—located about a hundred kilometers northeast of Delhi on the banks of the Ganga. The story of the *Mahabharata* probably had its origins in a territorial conflict between Indo-Aryan tribes living on the Gangetic plain during the first millennium B.C. Over time, accounts of these battles have been expanded through the interpolation of different myths into an overriding narrative of Hindu culture. In the *Mahabharata* the five Pandava brothers are the righteous heroes and the Kauravas are generally regarded as villains, though these distinctions are never quite so clearly drawn and moral ambiguity is a constant theme. During this struggle, in which the gods themselves participated, one of the decisive events involved a war elephant called Aswatthama, who shared his name with the son of Dronacharya, commander of the Kaurava armies. At a point where the battle had turned against the Pandavas, their divine charioteer, Krishna, suggested that the only way they could defeat the Kauravas was to announce that they had killed Aswatthama. Krishna explained that if the Kaurava commander heard news of his son's death, he would throw down his weapons in grief. The Pandavas were shocked that Krishna, an avatar of Vishnu, would suggest such deceit and Arjuna, the bravest of the brothers, refused to repeat this lie. However, his brother Bhima, who was the strongest of all, solved the dilemma by clubbing the elephant Aswatthama to death. After this, Yudhisthira, the eldest Pandava, accepted responsibility for their treachery and proclaimed that Aswatthama had been killed. (Unable to utter a complete lie, however, he added that the victim was "Aswatthama the elephant," though this part of the message was lost in the din of battle.) Once Dronacharya believed his son was dead, he lost all desire to fight and allowed himself to be killed. The Pandavas quickly turned the tide of the battle and after their victory over the Kauravas they regained power in Hastinapur.

Though the unfortunate elephant in this epic plays a minor role compared to the dramatic actions and moral dilemmas of the human characters, his presence in the story is significant. The *Mahabharata* is essentially a legend of kingship in which the Pandava heroes do battle to restore their claim to the throne. The fact that they must slay an elephant in order to gain power raises ethical questions not only about their deceit but also about the symbol of royalty that they destroy. An interesting comparison can be made with the *Ramayana,* in which a kingdom is lost because of the arrow fired at what was presumed to be an elephant—another fatal case of mistaken identity.

Historians who have chronicled the ebb and flow of India's many empires and dynasties faithfully record the number of elephants possessed by different kings as an accepted measure of the power and stature of each ruler. But counting elephants has never been particularly accurate and these figures are always prone to exaggeration. The number of elephants in Indian armies varies wildly from one account to the next. According to Greek sources, Porus had anywhere from 82 to 200 elephants in his army when he fought against Alexander. Farrukhi's verses imply there were 1,700 under the command of his patron Masud, son of Mahmud of Ghazni, the first Muslim invader to hold power in India. The Mughal emperor Akbar, at the end of the sixteenth century, was said to have amassed 2,000 elephants in his stables (some sources say 30,000), though only 500 were used in battle. A hundred years earlier, when Timur's Mongol armies swept down from Samarkhand and sacked Delhi, they faced a depleted force of only 120 war elephants.

Regardless of the exact numbers on a battlefield, elephants were obviously an indomitable presence at a time of war and inspired a sense of awe and fear. Amir Khusrav, a Sufi poet who was born at the end of the thirteenth century, provides a vivid image of the war elephants of Delhi:

> *The rank of elephants was like a line of baneful clouds,*
> *Each cloud with lightning to attack, swift like the wind,*
> *In its swift motion each elephant like a splendid mountain,*
> *The armour upon it like the cloud upon the mountain.*

The metaphors Amir Khusrav employs for the elephant are exactly the same as those that Kalidasa used a millennium earlier, though one describes the ferocity of battle while the other evokes the passions of love. Unlike Kalidasa, who composed his verses in Sanskrit, Khusrav was writing in Persian. He was one of the first poets of his time to use the language of invaders with a sensibility rooted in India. By echoing the metaphors of Kalidasa and other poets before him, Amir Khusrav carries on a lyrical tradition that celebrates the elephant in verse. But regardless of poetic imagery, war is never romantic, and Khusrav portrays the same elephant after the fighting has ceased.

> *From arrows the elephant was grafted with arrow-notches*
> *Like a porcupine with its back full of quills;*
> *From the elephant its driver was hanging,*
> *His body hanging and his life fled.*

Historians have examined the tactical value of elephants on the battlefield, citing examples of how these animals often stampeded when frightened or wounded, killing and injuring more of their own troops than the enemy's. A. L. Basham in *The Wonder That Was India* writes: "The great reliance placed on elephants by Indian tacticians was, from the practical point of view, unfortunate. . . . Even the best trained elephant was demoralized comparatively easily, especially by fire, and when overcome by panic it would infect its fellows, until a whole squadron of elephants trumpeting in terror, would turn from the battle, throw its riders, and trample the troops of its own side."

One of the most informative works on this subject is *War-Horse and Elephant in the Delhi Sultanate,* by Simon Digby. His monograph is erudite and eccentric, chock-full of obscure statistics drawn mostly from Persian sources. Digby is able to convey the imposing presence of elephants on a battlefield, as well as the manner in which these animals captured the imagination of kings and conquerors. He writes: "The elephant is a picturesque animal, and medieval authors are all convinced that it was a great asset in battle. Examples of its performance on the battlefield during the Sultanate period do not decisively support this view, and we must allow for the aesthetic enthusiasm which the elephant, like the horse, evoked."

Though he shares some skepticism about their strategic efficacy, Digby takes issue with Basham's complete dismissal of elephants as an asset in battle. He goes on to argue that during the period of the Delhi Sultanate (A.D. 1192–1398) strength and success on the battlefield were often measured by the quantity and quality of elephants deployed by an army. He also makes a case for their symbolic importance, demonstrating how the constantly shifting and often unpredictable power of Delhi's sultans depended to a very large degree on their control over the pil-khanna, or "elephant stables." In one dramatic instance, Digby describes how Nusrat Shah, a pretender to the throne of Delhi in 1393, was able to rally support after mounting one of the sultan's elephants during an insurrection. Following a great deal of intrigue and bloodshed, Nusrat Shah's principal ally, Mallu Khan, turned against him and usurped the elephant stables. As holder of the pil-khanna he was declared "the possessor of power and pomp." Shortly afterward, Mallu Khan reinstated the sultan on his throne, "with the prestige of all the elephants."

One of the most important questions is where these animals came from. Unlike most domesticated animals, elephants seldom breed in captivity. It also takes them over twenty years to reach maturity, which means that raising young calves was an expensive investment. Most rulers in ancient India never reigned long enough to have seen the returns. As a result, the majority of elephants were captured in the wild. By the time of the Delhi Sultanate, the forests in northern and central India had been severely depleted, and the supply of elephants would have come from remote areas, often outside the control of a sultan. Bengal continued to supply many of the elephants, and Digby's research indicates that Gujarat and Rajasthan, southwest of Delhi, still held scattered elephant herds in the fifteenth century, though these soon disappeared.

Aside from the difficulty of supply, proper training could take years before an elephant was ready to assume its position on the battlefield. Many were unsuitable for war, either because of their size and strength, or disposition. By far the most efficient means of acquiring war elephants was to capture them from an enemy. Military success in ancient India usually meant that the victorious king was able to take control of his opponent's elephants. It was primarily in this manner that emperors from Chandragupta to Akbar were able to accumulate as many animals as they did. Like ivory chess pieces carved in the shapes of tuskers, the war elephants themselves were strategic tokens of victory. Similarly, when empires and kingdoms began to crumble, peripheral governors and feudal tributaries asserted their independence by taking control of elephants stabled in outlying provinces.

The logistics of maintaining a pil-khanna would have been almost as overwhelming as the elephants themselves. An experienced mahout, as well as two or three fodder cutters, was the minimum retinue required for each elephant. If Akbar actually had 500 war elephants and the average daily consumption of fodder for each animal was 200 kilos, 100 tons of grass and leaves had to be supplied each day. The sheer volume of biomass is staggering, not to mention the quantity of dung that had to be disposed of, which makes it hard to imagine how these numbers could have been sustained. Perhaps more important than food was water. Even though the Yamuna River flowed through the imperial cities of Delhi and Agra (and was undoubtedly cleaner than it is today), when an army was on the march, it would have been a challenge to keep the elephants adequately watered and bathed.

In 1571 Akbar ordered construction of a completely new sandstone fortress at Fatehpur Sikri, which remains one of the most impressive feats of Mughal architecture. Among several grand entrances to this fort and palace was the hathipol, or "elephant gateway," with two carved tuskers, their trunks raised overhead to form an arch. Akbar also constructed a monument in memory of one of his favorite elephants. For all its extravagance, however, Fatehpur Sikri was ultimately abandoned. Some historians have speculated the cause was a shortage of water. Though a large tank was built to store rainwater during the monsoon, it proved inadequate. One can easily imagine that part of the reason the tank at Fatehpur Sikri ran dry was the emperor's elephants, who must have siphoned off thousands of buckets of water each day.

Despite the enormous expense, elephants remained an important part of military operations. If they were simply a symbol of prestige, it is unlikely that rulers would have employed them with such consistency. More important than their role on the battlefield was their ability to transport supplies. The elephant is particularly suited for traveling over rough terrain, through dense jungles and wetlands. Feroz Shah Tughluq, one of the Delhi sultans, used his elephants in an ingenious manner to cross a flooded river. As Digby describes: "The elephants were deployed in two chains, attached to one another by ropes, and were made to stand upstream and downstream of the ford. The chain of elephants upstream broke the force of the rushing waters, and the chain downstream served as a net, against which horses and riders who had been carried off by the water were caught instead of being swept away altogether." During a siege, elephants were also used as battering rams, to break down the gates of a citadel. Most of India's forts have metal spikes on the doors to discourage this kind of attack.

Instead of invading across the Hindu Kush, as earlier conquerors had done, ships belonging to the London merchants of Leadenhall Street dropped anchor at the ports of Calcutta,

Madras, and Bombay in the latter part of the seventeenth century. Bengal became the center of the East India Company's trading monopoly, dealing in cotton, leather, indigo, tea, and spices. Military adventures accompanied these conquests, but it was profits more than swordplay that drove the empire. After the rebellion of 1857, the British government officially took over power from the East India Company and Victoria was crowned empress of India. In exchange for their support, the maharajahs and nawabs were allowed to rule over their subjects and maintain a royal lifestyle, which included the ceremonial presence of elephants.

In 1903, soon after Victoria's death and the ascension of Edward VII, India's viceroy, Lord Curzon, organized the Delhi Durbar in honor of the new king emperor. An impulsive, ambitious, and egotistical man, Curzon embodied all of the romance and pragmatism of the Victorian era. One of the many contradictions of the British Raj was the overt romanticism with which the history and traditions of ancient India were celebrated by colonial administrators, while at the same time they expressed a patronizing contempt for the people over whom they ruled. Whether it was orientalist scholars deciphering the nuances of Sanskrit verse and revealing the wisdom of Hindu philosophy, or a pigsticking cavalry officer who stumbled upon the cave temples at Ajanta, there was a pervasive sense among Anglo-Indians that only they could locate, preserve, and redeem the wonders of Hindustan.

The choice of Delhi for the Coronation Durbar underscored Curzon's desire to assume the mantle of past empires, and the Mughal fortress afforded an appropriate backdrop for imperial succession. The elephant was to play an important role in this pageant. Of all the symbols of India, *Elephas maximus* provided one of the most compelling images of dominion, as Kipling's verses suggest:

The torn boughs trailing o'er the tusks aslant,
The saplings reeling in the path he trod,
Declares his might—our lord the Elephant,
Chief of the ways of God.

The black bulk heaving where the oxen pant,
The bowed head toiling where the guns careen,
Declare our might—our slave the Elephant,
And servant of the queen.

In his introduction to *The Great Indian Elephant Book,* D. K. Lahiri-Choudhury states, "The use of elephants by the British in state processions was perhaps the most important visual symbol of the process of orientalization of the Raj." Edward VII had toured India in 1876, as Prince of Wales, and he had already been carried in procession on the backs of elephants in princely states like Jaipur. He had also been taken tiger shooting on elephant back, drawings of which had appeared in *The Illustrated London News,* fueling the imagination of the British public with dramatic scenes of hundreds of tuskers lined up to beat their way through the jungle. Though Edward VII did not attend the Delhi Durbar, he was represented by the viceroy as well as the duke and duchess of Connaught, who were seated on a reviewing stand in front of which the procession passed. As one observer, Dorothy Menpes, described it: "For centuries the Oriental idea of power has been connected with superb shows and ceremonies. . . . The very fact that in a moment of time a Western Ruler should come and beat them at their own ground has established in the mind of prince and peasant alike a lasting impression of the position of the British Raj which nothing else could possibly have achieved."

From all accounts, the elephants were the most impressive part of the celebration and their images were relayed around the world. Curzon himself arrived at the Durbar on a tusker, the viceroy seated beneath a "glittering gold umbrella." The splendor and pageantry of the Delhi Durbar was projected as far away as America, where the combined Ringling Brothers and Barnum and Bailey Circuses re-created the coronation procession with their elephants.

Attuned as he was to symbols of history, Curzon understood the enduring significance of elephants, their dignity and power, as well as their longevity. The Delhi Durbar was choreographed to reassert British dominion in India, for the first stirrings of the independence movement had already begun. Just a decade earlier, in 1893, Lokmanya Tilak had organized the Ganesha Chathurthi festival in Bombay as a means of circumventing rules against public gatherings. Rather than allow the British to be perceived as a detached foreign power exploiting India purely for its own gain, Curzon attempted to present an image of legitimate and sustained control. Like Akbar, who was believed to have established power by integrating himself into the culture, Curzon saw the British Raj as the protector of India's past grandeur, which could only be reaffirmed through a modern European presence. In this context, the elephants with their maharajahs were paraded in front of the world to show that British India was firmly supported on the backs of these regal beasts.

The only person who didn't seem to understand the symbolism was Lord Kitchener, who was at odds with Curzon and insisted on riding in the Durbar procession mounted on a thoroughbred racehorse. The high-strung Derby runner took one look at the elephants and nearly threw the commander in chief out of his saddle. Kitchener was barely able to control the horse and rode sideways for the entire procession.

Throughout the nineteenth century British military forces in India made use of the elephant, primarily for hauling artillery pieces and supplies. By this time, guns and cannons had rendered most war elephants obsolete. However, elephants did serve to further the territorial designs of the British empire when Colonel

George Everest and others used them to transport heavy equipment through uncharted jungles in the Great Trigonometrical Survey of India.

During World War II elephants were employed by the British Fourteenth Army in northeast India and Burma to fight the Japanese. As Field Marshall Sir William Slim put it, "Without them our retreat from Burma would have been even more arduous and our advance to its liberation slower and more difficult." Under the command of Colonel J. H. Williams, these elephants and their handlers proved useful for jungle warfare, though their primary contribution remained the transport of supplies and the building of bridges, often under heavy fire. The majority of these elephants were originally owned by the Bombay and Burma Trading Corporation and were trained for hauling teak, a skill that was invaluable in constructing bridges. During the course of the war the elephants suffered severe casualties from bombs and land mines. In his book, *Elephant Bill,* Williams describes how these tuskers became expert sappers and helped construct the military road from India into Burma. One of the many anecdotes that Williams recounts is a confrontation between an army bulldozer and a tusker, neither of which had ever seen the other before. After initial hesitance on the part of the elephant, the two were soon working side by side.

Orwell's Dilemma

One of the greatest essays in the English language is George Orwell's "Shooting an Elephant." On the surface it is a simple, straightforward account of an experience the author had when he was a young police officer in Burma. However, as Orwell tells the story, it becomes a condemnation of the British Empire. Born in India, Orwell understood firsthand the complexities and contradictions of the Raj. He writes with the clarity and passion of someone who was intimately involved in the colonial experience.

And in the end he holds himself as well as his fellow Anglo-Indians accountable for the oppression and injustices of British rule.

The elephant in this story was not one of the royal tuskers at the Delhi Durbar, nor was it a battle-hardened bull from the Mughal pil-khannas. It was an ordinary working elephant, probably used for hauling logs or pulling automobiles out of the mud. As Orwell describes it, the elephant was in musth and during a fit of rage broke free of its chains. We are told that the owner was an Indian, most likely a merchant who settled in Burma during the Raj, as others did in colonies throughout Southeast Asia, Africa, and the West Indies. His elephant had leveled a bamboo hut, gored a cow, destroyed several fruit stalls, and overturned a municipal rubbish van. But its greatest crime was trampling a man to death—coincidentally another Indian, one of many laborers who also came to Burma in the footsteps of the British.

Summoned to "do something about" this musth elephant, the young police officer set out alone through the town, carrying his rifle. As a gathering crowd of Burmese spectators followed, he was uncomfortably aware of their desire to see the elephant killed. When he finally came upon the animal, now calmly feeding in a paddy field, Orwell initially resisted the idea of shooting the elephant. Yet he realized that the throng of people watching him all believed he would use his rifle. If he didn't, they would consider him either a coward or a fool.

> And it was at this moment, as I stood there with the rifle in my hands, that I first grasped the hollowness, the futility of the white man's dominion in the East. Here was I, the white man with his gun, standing in front of the unarmed native crowd—seemingly the leading actor of the piece; but in reality I was only an absurd puppet pushed to and fro by the will of those yellow faces behind. I perceived in

this moment that when the white man turns tyrant it is his own freedom that he destroys.

Submitting to the expectations of the crowd, Orwell reluctantly lay down on the ground, took aim, and fired. After a second shot the elephant "sagged flabbily to his knees. His mouth slobbered. An enormous senility seemed to have settled upon him. One could have imagined him thousands of years old. . . . He was dying, very slowly and in great agony, but in some world remote from me where not even a bullet could damage him further."

The essay ends with a debate that took place within the European community. Some felt that Orwell had "done the right thing" and that he was legally obliged to shoot the tusker. Others believed that the destruction of a working elephant, whose behavior would have returned to normal as soon as its musth wore off, could hardly be measured against the death of an insignificant "coolie." Orwell himself found these arguments as hollow as his own motives.

"I often wondered," he concludes, "whether any of the others grasped that I had done it solely to avoid looking a fool."

Orwell is not a writer who dwells on metaphors and much of his work exposes the lies behind symbolic images and gestures. The arrogance of white sahibs carrying guns is one of the obvious targets in "Shooting an Elephant." A revealing comparison can be made between Orwell's essay and shikar stories that were a prevalent and popular feature of literature during the Raj. Whenever memoirs of India were written by military officers, civil servants, tea planters, and other colonials, hunting was a requisite feature—usually a story or two that involved tigers and elephants.

One of the most celebrated white hunters during the nineteenth century was a man named Roualeyn Gordon-Cumming who was described as "the greatest hunter of modern times." He has the dubious distinction of being quoted by Charles Darwin on the subject of elephants' weeping, as Gordon-Cumming describes how "tears trickled" from the eyes of a wounded tusker he had shot. Roualeyn's younger brother, William Gordon-Cumming, was also a well-known shikari in India, publishing a book called *Wild Beasts and Wild Men*. More than anyone, these two brothers represented the brute arrogance that Orwell despised.

There is no direct evidence that Orwell knew anything about William Gordon-Cumming or that he had read of his exploits. Yet the "The Terror of Hunsar," a shikar story in which Gordon-Cumming is the lead actor of the piece, bears uncanny similarities to "Shooting an Elephant." Though the events described date back to the 1870s, this account was retold by A. Mervyn Smith in Calcutta's *The Statesman* newspaper around the turn of the century.

In this case the elephant has a name, Peer Bux, and he is described as standing nine-and-a-half feet at the shoulder, with three-foot tusks. He belonged to the Madras Commissariat and was prized for his ability to haul gun carriages out of the mud and carry loads of a ton and a half. Except for periodic bouts of musth, Peer Bux was generally well-behaved and never threatened or harmed his handlers. After six years of service, however, he broke his chains and went on a rampage. (His mahout, Smith tells us, had gone off with G. P. Sanderson to Assam on a khedah operation.) Escaping into the forest, Peer Bux continued his attacks and was eventually declared a rogue.

As the tusker was greatly valued by the Madras government, there was reluctance to have him shot, even though he had killed a number of villagers and destroyed private and government property. Once his death sentence had been passed, several hunters made unsuccessful attempts to destroy Peer Bux and one even lost his own life. Finally, Gordon-Cumming agreed to take up the challenge and tracked Peer Bux down to the banks of the Kabini River.

A. Mervyn Smith narrates this fatal encounter with all of the flatulent bombast of empire that Orwell sought to expose. Here is the white man with his rifle, but instead of an expectant throng of "natives" at his back, he is accompanied only by his tribal tracker, Yalloo. The two of them sit together on the riverbank and Gordon-Cumming calmly eats a biscuit as he waits for Peer Bux to appear. Soon enough the rogue arrives, trumpeting loudly and charging down the river bank. Unable to get a clear shot, Gordon-Cumming waits until the elephant is a few yards away, then casually tosses his pith helmet on the ground. Peer Bux stops for a moment and lowers his head to sniff the topee, at which point the hunter puts a bullet in his brain.

Yalloo the tracker is given the last word (heavily paraphrased by A. Mervyn Smith). He regales us with Gordon-Cumming's heroics, "Ah, comrade. I could have kissed the Bahadoor's [my lord's] feet when I saw him put the gun down, and go on eating his biscuit. . . . I was ready to die of fright; yet here was the *sahib* sitting down as if his life had not been in frightful jeopardy just a moment before. Truly, the *sahibs* are great!"

This is exactly the kind of self-congratulation that Orwell attacks in his essay, the racial arrogance and false heroics of the empire. Growing up in India, it is likely that Orwell heard many hunting yarns like this. Perhaps he even read A. Mervyn Smith's account in the Calcutta papers or found a copy of *Wild Beasts and Wild Men* on a bookshelf at the British Club in Rangoon. While these shikar stories may have inspired others to emulate Gordon-Cumming's adventures, someone as perceptive as Orwell could only have recognized the lie.

Project Elephant

Despite its imperial history and grandeur, Delhi today is essentially a city of bureaucrats. Politicians may posture and make speeches, but much of the real power of government lies in a convoluted administrative structure of ministerial commissions, corporations, and advisory boards that are controlled by secretaries, joint secretaries, additional joint secretaries, and others on down the line. Entering the Central Government Office Complex off Lodhi Road is like stepping into a concrete labyrinth. The clustered ranks of multistory towers look as if they were designed by committees of civil servants who confused their organizational charts for architectural blueprints. Checkered lines of windows ascend like a rigid hierarchy while signs point in all directions but give little away. After some searching I found Paryavaran Bhavan—Environment House—and took an elevator up to the third floor. The only reassuring aspect of this building was a series of wildlife posters on the walls. In many ways, the fate of India's elephants depends as much on these corridors of power as it does on forests and grasslands.

S. S. Bist, director of Project Elephant, is refreshingly unbureaucratic. A trim, animated man, who holds the rank of Inspector General of Forests, he began his career in the West Bengal Forest Service and seems more suited to the simpler surroundings of a wildlife sanctuary than the high-rise strictures of the office complex.

"I have spent most of my career dealing with elephants," he told me. "In West Bengal we have a wild population of only 300 but these are intensely managed. Half of the time I was chasing elephants and the other half they were chasing me."

Project Elephant was started in 1992 in response to widespread concern over the future of the Asian elephant and destruction of its habitat. According to Bist, there are approximately 50,000 wild elephants in Asia and 16,000 in captivity. India has the largest population in the region, around 28,000 elephants in the wild and 3,500 captive animals. Figures like these are always difficult to verify, but compared to the population of African elephants, estimated to be above 300,000, *Elephas maximus* is clearly outnumbered.

Like Project Tiger, which began in 1973, the strategy of Project Elephant is to focus conservation efforts on a single major species in order to preserve its habitat and guard against poaching. By extension, these efforts help protect and sustain other fauna and flora that share the same environment. Project Tiger has had some success in bringing the tiger back from the brink of extinction. Whether it will ultimately save the species remains to be seen but a precarious level of stability has been achieved.

"What we are doing is quite different from Project Tiger," said Bist. "Rather than calling elephants a 'flagship species' I think of them as a 'stand-alone' species." The reason for this, he argues, is that unlike tigers, who are largely contained within national parks and tiger reserves, the distribution of elephants is much more scattered. "Only 23 percent are found in national parks. The rest range across other reserve forests and private lands."

As a central government agency, Project Elephant assists forest departments in various states by providing funding and technical expertise for wildlife management. It also supports scientific research. One of the primary objectives is to restore and protect the natural habitat of elephants as well as forest corridors that link isolated populations. Aiding state efforts to guard against poaching and pressing for action against illegal ivory smuggling are other priorities. Project Elephant also provides financial compensation to farmers affected by crop raiding and to the families of those who have been killed by wild elephants.

"Of approximately 350 deaths attributed to wild animals each year, at least 300 are caused by elephants," said Bist. "They are also responsible for much of the damage to crops, especially in places like Assam and Karnataka."

Though the hunting and capture of wild elephants is banned in India, Bist and others feel that it still offers one of the best solutions to human-elephant conflict, while helping to preserve the species. In Assam and West Bengal thirty-three "problem elephants" were recently captured by the forest departments of those two states, in an effort to protect both the elephants and human beings. As Bist has written, "Domesticated elephants provide a sort of insurance against extinction of their wild brethren . . . domestication of elephants is now a conservation imperative." Though breeding in captivity remains a challenge, the adaptability of the species does lend itself to maintaining elephants in forest camps like Theppakkadu, where females occasionally mate with wild bulls. If habitual crop raiders can be captured and kept inside the boundaries of national parks and sanctuaries, these populations can serve as a genetic bridge between tame animals and those that remain in the wild. However, Bist and other experts have pointed out that the skills and knowledge required for elephant catching and handling are rapidly dying out and need to be revived. With fewer and fewer elephants in captivity, there are fewer mahouts to carry on these traditions.

Obviously, there is room for debate over the ethics of capturing wild elephants, but it does offer a viable alternative to the present state of affairs, when crop-raiding or dangerous elephants are often shot or poisoned by farmers. The culling of elephants in India, particularly when they are a threatened species, seems both unnecessary and unwise. Capture and relocation, whenever feasible, would appear a much more logical option, though it is not without problems. Wild elephants can suffer injuries when captured, and there is always the potential for abuse from poorly trained and insensitive forest department officials and mahouts.

One of Bist's concerns is the proper treatment and monitoring of captive elephants. He recently edited and republished A. J. W. Milroy's book, *Management of Elephants in Captivity,* which advocates a humane approach to capture and training. Project Elephant supports veterinary camps at the Sonepur Mela and in Jaipur and Delhi. Through state forest departments it also funds workshops and training sessions for mahouts to promote the use of modern medicine and more humane methods of handling elephants. Enforcement of established standards for animal welfare and prevention of cruelty to elephants also comes within its purview, but Project Elephant's main challenge is encouraging state governments to move beyond policy to implementation.

Based on his experiences in West Bengal, Bist feels that many other state forest departments could utilize their trained elephants more effectively. In fact, he feels that employing forest department elephants for tourist rides in places like Corbett Park is a mistake and that private contractors should be used instead. This would provide much needed employment for mahouts and their tame elephants, while freeing up forest department elephants and personnel to carry out their official duties.

"Forest officers must be encouraged to use elephants for patrolling, rather than jeeps and motorcycles. It is quieter and easier to reach those areas where poachers hide."

Even though he is part of the bureaucracy of the Ministry of Environment and Forests, Bist conveys a genuine passion for elephants, which is even reflected in part of his e-mail address—"Gajendra." He also happens to be married to one of India's chief experts on elephant handling and training, Parbati Barua, whom he first met when he was a divisional forest officer in West Bengal. The two of them share a lifelong commitment to the survival of this species.

"I owe everything to elephants," said Bist, with an expansive gesture. "Even my wife!"

northeast

The University of Nature

*T*wo thousand kilometers east of Delhi lies Guwahati, the capital of Assam and "Gateway of the Northeast." Once an isolated betel nut bazaar, Guwahati is now a modern city, with all of the amenities and aggravations of the twenty-first century. At the airport I was met by Hemanta Das, a wildlife guide who handled my travel arrangements in Assam. He also helped me find the way to Parbati Barua's house. Diversions and detours make it difficult to navigate Guwahati's streets, especially in the dark. In addition, the directions we had been given were cryptic, with landmarks like an unnamed roadside café and a broken-down bus, but eventually we found her.

A slight, pensive woman with her hair pulled back in a tight bun, she was wearing a hand-block-printed sari. We were ushered into a drawing room with family pictures on the walls and the heads of deer and gaur glowering down from above. My first impression of Parbati Barua was that she had little patience for small talk.

"The first time you rode on an elephant, how old were you?" I asked.

"One year and seventeen days," said Barua, sitting up straight in her chair as if mounted on an elephant.

Her father, Laljee Barua, was maharajah of Gauripur, famous as a hunter and as an elephant handler. In his lifetime, he captured a total of 1,029 elephants, many of which he trained himself. Following in her father's footsteps, Parbati Barua has studied and worked with these animals for most of her life. Sometimes referred to as "the only woman mahout in India"—a totally inadequate description—Barua has been the subject of a documentary film and a book called *Elephant Queen*. She is something of a celebrity in Assam. With two trained elephants, or kunkies, of her own, Barua is often called on to assist the forest department when wild herds are causing trouble—either by driving them away from fields or capturing them for relocation.

"Capture can be a part of conservation," she said, explaining that the following week she was going to Chattisgarh, in central India, to help control wild elephants that were raiding crops.

"Elephant-human conflict is not the fault of elephants," she said. "It is the fault of man. Human beings have encroached on the elephant's habitat."

When asked about the relationship between a mahout and an elephant, she said, "It is like family—brother and sister, husband and wife."

"But do the elephants think of the mahout that way?"

"Of course," she said, decisively. "If you give any animal love and care, it will respond in kind. The only creatures that don't do this are Homo sapiens."

"Most mahouts are men. Is there a difference because you're a woman?"

"No. It doesn't matter if I am a woman or a man. When I am with my elephants, I am an animal."

Earlier I had been told that Barua and her elder sister, Pratima Pandey, are known as hastirkanya—daughters of the elephant. Pratima Pandey, who died recently, was a well-known folksinger and many of her songs are based on the lore of elephants and their handlers. In Assam, romance surrounds the mahout because he is able to tame such a large animal as an elephant and he goes into the forest alone and unafraid.

"He is a hero!" said Parbati Barua. "Young girls fall in love with him when they see him riding on an elephant."

Folk songs celebrate the bravery of mahouts whose work in the forest often takes them away from their lovers. The lyrics in one of Pratima Pandey's songs lament this separation.

> O my mahout on a tusker.
> Can a boat sail in a lake without water?
> What can a woman do with her beauty
> If her mahout is no longer near her?
>
> O my mahout on a tusker.
> I left my mother and my brother
> I left my golden home
> And now you have left me, weeping.
>
> O heartless mahout on a tusker.

Other folk songs compare the dangers and challenges of catching wild elephants to the experience of love. Bupen Hazarika, another legendary Assamese singer, answers with the words of the mahout:

> O girl from Gauripur
> When I went to capture elephants
> I also fell in love with you.
> It is easier to control a wild elephant than this girl.

Parbati Barua understands the romance of elephants, but she speaks of them not only with passion but also with pragmatism. The proper care and training of elephants is something that she advocates in articles she has written, as well as in workshops that she holds for forest department mahouts. She also regularly attends the Sonepur Mela and has been active in establishing a humane approach to elephant management and training. In collaboration with S. S. Bist, she has published an article, "Cruelty to Elephants—A Legal and Practical View," which concludes that the laws in India governing the treatment of wild and domestic animals are adequate but that "suitable norms and standards for the management of captive elephants are yet to be fixed and given the backing of the law."

Though khedah operations used to be conducted in Assam many years ago, the most popular technique of catching elephants in the Northeast has always been mela shikar. It has been described as a more humane method, though not without danger to animal and man. Tame kunkies are ridden into the midst of a wild herd so that selected elephants can be lassoed with grass ropes. Only the most experienced mahouts are able to noose wild elephants and Barua explained that these skills are rarely practiced today.

"We need to preserve this traditional knowledge of elephant catching and training; otherwise it will be forgotten," she said.

Barua's approach and opinions are controversial. Some wildlife activists argue that using a tranquilizer gun is a more effective and humane method of subduing and capturing a wild elephant than mela shikar. Unfortunately, a few months after we met, Barua found herself accused of animal abuse, when a young male elephant that she captured in Chattisgarh died while being trained. She has defended herself by saying that the animal was already injured by tranquilizer darts fired by a forest department official. Whatever the facts may be, the problem of human-elephant conflict is not easily solved and requires new and innovative solutions as well as traditional techniques.

Unlike most elephant experts, Parbati Barua is not a trained zoologist, but she has firsthand knowledge of these animals and understands them on an instinctual level.

"A scientist looks at a flower and dissects it and tells you how many petals it has and all of its different parts," she said. "I look at the flower and recognize its beauty. For me, it is the same with elephants. You must spend time with these animals to understand them. A scientist will study their anatomy and take a mathematical approach. I am not able to tell you any of that, but just by looking at an elephant I know if it is healthy or not. I can also tell you where it came from—whether it is from the north bank or the south bank of the Brahmaputra. I don't know how I can tell you this, but I can."

She speaks in a clipped, no-nonsense manner, but underneath the sharpness of her voice is a sensitivity in her words. Near the end of our conversation, she looked at me with a stern expression and said, "In the University of Nature there is no syllabus. Each day I learn something new."

Having spent most of her life in the company of elephants and their handlers, Barua has written about the image and mystique of the mahout in northeastern India. "With their capacity to control such a big and powerful animal as the elephant, mahouts are often associated with supernatural powers and invited to act as 'faith healers' or 'Ojhas.' Some of them practice witchcraft. Many of the mahouts have made a name for their knowledge of medicinal herbs."

The former princely state of Gauripur, Barua's ancestral home, lies in Goalpura District to the west of Guwahati. Located at a point where the Brahmaputra River turns southward into Bangladesh, this region has always been famous for elephants, though most of the forests have now been cleared for cultivation. Immediately to the south of Goalpura District are the Garo Hills, in the state of Meghalaya—home of the clouds.

Originally part of Assam, the Garo Hills are a tangled chain of ridges that taper down to the Brahmaputra. At one time these hills were full of wild elephants, though spreading agriculture and human depredation have reduced their numbers, squeezing the last remaining herds into a few pockets of protected forest like Balpakram National Park.

The villagers of the Garo Hills have a tribal culture that reflects their long association with the forests and an intimate knowledge of animals and plants. Though the majority of the population are now Christians, converted by British and American missionaries over the past two centuries, their folklore and beliefs come out of an animistic faith. A small minority of the Garo people still worship arboreal spirits and believe in a mystical connection between human beings and other animals. Even among those who have become Baptists or Catholics, many Garos hold on to remnants of these primal beliefs, which are older than any form of organized religion and still speak to those who remain connected to the forests and the land. Though these beliefs have often been dismissed as primitive superstitions, the underlying narratives are as relevant today as the stories of other faiths. In fact, as the environment out of which they come is rapidly destroyed, the message in this mythology becomes all the more urgent.

One of the essential aspects of Garo folklore is the kinship between human beings and animals. There is a belief that the human soul can travel outside the body and inhabit another creature, through a form of "psychic transmission." Dewansing Ronmitu Sangma, a Garo folklorist, has compiled many of these oral legends in a book called *Jadoreng,* which explores the spirit world of the Garo Hills. Two of these folktales tell the story of man and elephant.

Many years ago, when the British still ruled India, there lived a shaman named Ganjang Marak Napak. He was able to project his soul into the body of a rogue tusker who had attacked and killed many people in the Garo Hills. While sitting at home with his wife and family, Ganjang would narrate all of the atrocities committed by this rogue, recounting in gruesome detail how he had crushed men to death with his feet and gored them with his tusks. Later, when news arrived of the tusker's latest rampage, all of the details in Ganjang's story were found to be true.

Finally, in June 1924, the deputy commissioner, G. D. Walker, ordered a contingent of the Assam Rifles to hunt down this rogue, who was responsible for the death of fifty-two people. The soldiers entered the forest and finding the elephant without much difficulty, they fired a volley of bullets that killed it instantly. At that very moment, Ganjang the shaman was having a feast in his house and entertaining his guests with rice beer. Just as he was about to take a drink from the gourd, he let out a gasp and fell facedown on the floor of his hut. When his companions rushed forward to help him, they found that Ganjang was dead.

Another shaman, named Kawak Sangma Dawa, lived in the village of Simsanggiri. He was also able to exchange souls with an elephant, a noble tusker who lived deep in the forests of the Garo Hills. For the entertainment of his friends and family, Kawak often recounted the antics and adventures of this elephant, how it bathed in streams or rivers and fed on the wild fruits of the jungle. He also described how the tusker had fallen in love with a beautiful female elephant, relating all of the intimate details of their courtship and mating.

Around this time, during the winter of 1916–17, the maharaja of Durgapur was given permission by the British authorities to capture elephants in the Garo Hills. He sent a party of hunters and mahouts with trained elephants into the jungle. Using the method of mela shikar, the hunters roped and captured a number of animals, including the tusker's pregnant mate. She and the other elephants were then taken away to Durgapur, where they were tied up in the maharajah's pil-khanna.

Though far away in his village, the shaman observed all of this through the tusker's eyes and he told his family what had happened, lamenting the separation of the two elephants. Kawak could feel the tusker's sorrow and he confided in his wife: "Out of irresistible longing for his beloved spouse, my *other-self* is following her down to Durgapur. My heart is about to break with yearning. My life is approaching its twilight."

Distraught and inconsolable, the tusker left the forests of the Garo Hills and tracked down the hunting party, pursuing them all the way to Durgapur. Arriving at the pil-khanna, he allowed himself to be captured so that he could be reunited with his lover. But as the tusker was being put in chains, he picked up a lump of earth with his trunk and swallowed it, a gesture that conveyed his unwillingness "to remain alive as a slave of man." After this, he refused to eat and standing beside his beloved mate, the tusker slowly starved himself to death.

All of this was described in detail by the shaman Kawak, who experienced the tusker's suffering as if it were his own. Sitting at home with his wife and family, he also grew weaker and weaker, then finally died. After his body had been cremated, some of Kawak's relatives wondered if he had been speaking the truth. They decided to travel down to Durgapur and see for themselves. On reaching the maharajah's pil-khanna, they found the tusker's rotting corpse and when they spoke with the mahouts, everything that Kawak had narrated was confirmed.

Kaziranga

Screens of tall grass surround a broad basin where monsoon floods have receded, leaving a kidney-shaped lake at the center. Waterbirds cover its surface, flocks of whistling teal and spotbills, bar-headed geese, and brahminy ducks with rusty plumage. An adjutant stork stands in the shallows, shoulders stooped and beak tucked in, as if waiting for a fish to swim between its legs. Pelicans are nesting in a grove of trees beyond the grass, the leaves and branches spattered with their white excrement, like a coating of limewash. Far off in the distance, across the lake, I can see wild buffalo and swamp deer grazing on the floodplain. Through my binoculars the landscape is foreshortened, the lenses magnifying but distorting the scene so that everything is compressed into a narrowed field of vision. The blue shadows of the eastern Himalayas, draped with a gauzy layer of clouds, seem to rise directly above the animals instead of being far away, across the Brahmaputra River and beyond the borders of Assam. After I lower the binoculars, it takes a moment for my eyes to adjust as the landscape shifts back into perspective and the birds, the animals, and the mountains retreat and diminish.

Hemanta nudges my arm and points to our left. A herd of wild elephants have just emerged from cover and stand about four hundred meters away. Their shifting gray shapes are partly hidden by a curtain of grass, which protrudes into the lake bed like the wings of a stage. The presence of elephants brings the contours of the landscape into focus. A few minutes earlier, when I scanned the same area, the ground seemed flat and featureless but now as the herd moves across it, I can see troughs and depressions below the flood line and higher spits of land where the grass extends toward the lake.

Lifting the binoculars to my eyes again, I see the elephants more clearly now, standing about thirty meters from the grass. There are six of them, including two calves, but a moment later three more cows emerge, followed by a young tusker. At first they seem to be heading for the water and I expect them to cross in front of us, but they stop and turn around.

Something about their behavior seems unusual. Rather than grazing with sociable complacency, as most herds do, these elephants appear agitated. All of their heads are raised and they group themselves together with the calves in the center. Clearly, they are protecting their young, though it is difficult to see what the threat might be. One of the females makes a mock charge toward the high grass in front of her, then backs away nervously, bumping into the others as she retreats.

At the same time I see a pair of swamp deer come out of the grass and move in close behind the herd, as if trying to join the circle. Swinging about aggressively, one elephant chases off the deer, shaking her head at them in annoyance and making a trumpeting squeal. The swamp deer retreat a short distance, then skittishly return.

Adjusting the focus on my binoculars, I strain to catch some sign of movement in the grass, but from where we are standing it is impossible to see what is frightening these animals. The young bull, whose tusks poke out like ivory pegs on either side of his trunk, finally makes a furtive charge. He lumbers forward, taking a dozen quick steps, then comes to an abrupt halt just short of the grass. At the same moment, I see a streak of orange fur, as a large animal bounds from one section of grass to the next.

"Tiger!" says Hemanta, loud enough to make me jump.

Keeping my binoculars trained on the spot where it vanished, I wait for the tiger to appear again but it is gone. Though the predator must have been stalking the elephant calves, or perhaps the swamp deer, it has decided to give up the chase. Several minutes later, when I glance back at the elephants, they have moved another fifty meters across the lake bed. The calves remain in the center of the herd, though the adults now seem less concerned, shambling away from the scene of their encounter. The swamp deer are still using the herd as cover and once again a cow tries to shoo them off.

By this time my eyes ache from peering through the binoculars. The excitement of seeing a tiger, even as briefly as we have done, leaves me with a sense of elation and adrenaline in my veins. Blinking to readjust my vision, I see the elephants drift toward the trees, where the pelicans are roosting. Off to one side stand the swamp deer, their tawny hides blending in with the grass. Though nothing seems to have changed, everything looks slightly different, as if the brief confrontation has brought the scene into sharper contrast. Only the ducks on the lake seem unaware of what has just happened, babbling among themselves like ignorant bystanders.

In documentary films about wildlife, scenes of predators attacking prey are often spliced together into a frenetic medley of pursuit, then repeated again and again in slow motion while set to music. By comparison, watching animals in their natural habitat usually involves hours of patience and when the action occurs, it is often over before we know it. Yet the satisfaction of witnessing a dramatic event in nature or catching sight of a rare species makes up for all the waiting. Equally important is the appreciation we gain for subtle nuances in a landscape. Those few moments of excitement become all the more meaningful when we take time to appreciate the setting in which they occur, the variegated foliage of the forest and the insects, birds, and other animals that play a supporting role in the ongoing drama of death and survival.

Kaziranga National Park lies at the center of Assam, along the southern bank of the Brahmaputra. It covers an area of 430 square kilometers and is divided into three separate ranges, the eastern, central, and western sections of the park. Most of the grasslands become flooded during the monsoon and are inaccessible, except by boat, from the end of April to the beginning of November. During this period, when the park is closed, animals are able to seek higher ground where the forest cover is heaviest. In certain years, when the floods have risen particularly high, animals are forced to retreat into the range of Karbi Hills outside the southern boundaries of the park.

One of the primary reasons Kaziranga has survived in its natural state is its extensive wetlands, which make it unsuitable for cultivation. Those villagers who live near the perimeter of the park are mostly fishermen belonging to an indigenous community known as the Mising tribe (pronounced "Mishing"). Their huts are built on bamboo stilts to keep them above the floods and they use dugout canoes to negotiate the waterways. Lakes and rivers within the park are closed to fishing, but just outside the boundaries of Kaziranga are many ponds and drainage channels. Though most of the fishermen now have nylon nets, I saw a number of Mising villagers using conical bamboo traps. What they caught were fingerlings, the size of small sardines. I also saw a group of women catching crabs with circular bamboo sieves into which they scooped up mud from the shallow channels.

Along the national highway, which borders the southern perimeter of the park, are fields and villages, as well as several tea estates. Wildlife corridors lead into the adjacent Panbari Reserve Forest and other protected areas, but during the monsoon there is an increased risk of poaching, which has always been a problem at Kaziranga. The great Indian one-horned rhinoceros is a primary target, as are the few remaining tuskers in Assam. But an even greater risk to the elephant population is deaths that occur when fields are raided and villagers try to protect their crops.

Nowhere in India is the conflict between elephants and human beings as widespread and desperate as in Assam. Though the people of this region venerate elephants as manifestations of Ganesha, the destruction of paddy fields and other crops has often led to vindictive retaliation. Gunshot wounds are the most common cause of fatalities but poisoning and electrocution also take a toll. By dangling a wire from overhead power lines, along a path or near a water hole, some villagers inflict high-voltage revenge on the herds that raid their crops. Vivek Menon, who has investigated recent elephant killings in Assam, tells a story that illustrates the animosity of farmers. Painted on the carcass of one of the poisoned elephants Menon found graffiti equating the animals with terrorists. The message in Assamese: *"Dhan Chor Bin Laden!"* ("Rice Thief, Bin Laden!")

One can understand the frustration and anger of subsistence farmers whose entire livelihood can be threatened in a single night by elephants raiding their crops. Yet the fact remains that human beings have encroached on elephant habitat and inflicted far greater damage than the animals themselves. Out of a total area of roughly 20,000 square kilometers of reserve forest in Assam, about 7,000 square kilometers are occupied by illegal settlers. In February 2002, the Supreme Court of India ordered the eviction of these squatters, and some of the encroaching settlements were razed to the ground. Forest department elephants were even used as bulldozers to tear down the bamboo huts. The situation is complicated by a violent history of separatist agitations in Assam and other parts of the Northeast. Though the United Liberation Front of Assam (ULFA) has negotiated a truce with the state and central governments, there is still a contested sense of entitlement to the land and hostility toward authorities like the forest department. Illegal firearms remain in the hands of former ULFA militants and these guns are often used for poaching. As the forest minister of Assam, Pradyut Bordolai, has stated: "There is a nexus between smugglers, militants, and the encroachers." Though Assam has recently returned to a degree of political normalcy and travel restrictions for foreigners have been eased, the situation remains tense, and along with many innocent people, wild animals get caught in the crossfire.

The forest guards who patrol the park live a difficult and dangerous existence. While visiting the eastern range of Kaziranga, we were accompanied by a guard named Muhammed Nasir Ahmed. He was about thirty but looked several years younger, his boyish features partly hidden beneath the brim of a camouflage cap. Nasir was armed with a .315 bolt-action rifle, now standard issue for most forest guards in India, replacing the antique .303 rifles that were used until a few years ago and dated back to the First World War. In 1993 Nasir was attacked by a tiger and almost killed. With a hesitant smile, he recounted the story:

"I was on patrol with another guard. We were walking through the forest and there was a tree that had fallen across the path. When I started to cut the branches to clear it away, a tiger jumped on me and tried to carry me off. The other guard fired sixteen bullets into the air before the tiger finally let go. By then I was unconscious but most of the time I was aware of what was happening. Fortunately, my companion was able to contact our range headquarters by wireless and they sent a jeep to rescue me. I had lost a lot of blood and spent three months in the hospital before I could walk."

Rolling up his left sleeve, Nasir showed me the scars of a dozen puncture wounds that looked like patches on an inner tube. He said that when the divisional forest officer came to investigate the scene of the attack, the same tiger jumped on his jeep before running away.

Much of the eastern range of Kaziranga is densely forested. Though many of the trees are the same as those in other parts of India, the wetland jungles of the Brahmaputra valley have a distinctly different appearance. One of the characteristics is cane creepers that form a snarled mass of pleated leaves and serpentine stems. Driving in an open jeep, we often had to dodge their thorny tendrils that reached down from above like barbed hooks. Much of this jungle is impossible to enter on foot and each year, following the monsoon, trails have to be cleared and bridges rebuilt.

Forest camps are located at intervals of ten kilometers and each range of the park is divided into separate beats patrolled by teams of two or three guards. In this way, the entire forest is monitored, though much of it remains inaccessible because of undergrowth and marshland. The tall elephant grass, which covers most of the lowlands, is impenetrable, not only because it is so thick and tall but because the edge of each blade is sharp enough to slice a person's hand.

One of the common trees in Kaziranga is the elephant apple, or o'tenga, which can grow up to twenty feet high, with dark green foliage. The fruit that gives this tree its name is the size of a small melon with a leathery, green skin like that of an avocado. When ripe, it turns orange. Villagers in Assam eat elephant apples—sometimes as a curried vegetable but also as a sweet, mixed with raw cane sugar. The elephants, of course, eat them plain. Near many of these trees we found the half-chewed remains of unripe fruit.

The jeep road through the eastern range of the park ends at Ahotgiri Camp, where forest guards live in a one-room hut built six feet off the ground. When we arrived, they were making tea and invited us in to have a cup. Though the hut had a tin roof, the walls were nothing more than bamboo matting. Three beds, each with a mosquito net, took up most of the space inside. At the back was an open porch that served as a kitchen.

Subin Burman was the senior of the two guards, with a tweaked moustache and a melodramatic style of speech. He told us that because the moon was full these days, they had to stay awake all night and remain alert for poachers. Burman claimed to have caught several poachers in the fifteen years he has worked at the park. He also said that he has been chased by elephants many times and told us about a lone makhna that lived near their camp. Recently it charged him while he was on patrol and the elephant only backed off after he emptied a rifle clip in the air. Except in an extreme emergency forest guards would never fire at the animal itself. The other guard, Prabin Kakati, was a quieter man who had recently been transferred to the park from a social forestry project in Naogaon. A third guard was also assigned to Ahotgiri but he was on leave. Their only means of communication with range headquarters was a walkie-talkie, its batteries charged by a solar panel propped against a tree stump. The camp has no electricity or running water, though there is a marsh nearby where they can fill buckets. Next to their hut the guards tend a small vegetable patch, but Kakati said that deer and boar eat most of what they grow. A forest department jeep delivers rations once a week and they collect wild fruit and vegetables in the forest.

While we were having tea, a spotted dove flew into the hut and landed near our feet. It was perfectly tame and picked its way confidently across the floorboards, as Kakati sprinkled a few grains of rice for it to eat. The bird's delicate head bobbed up and down as it pecked at the rice. Burman told us that they used to have a pet cat to control rodents that came into the hut at night but it had been killed by a python. He performed a convincing pantomime of the cat being squeezed to death, wrung out like a piece of laundry, then swallowed whole.

"When I saw it happening," he said, "my first instinct was to kill the python, but then I thought, *This is how a snake feeds itself.* What could I do? The cat was already dead, so I let the python go."

Later that day, in another section of the park, we saw a python by the side of the road. It was lying in the sun, digesting a meal, and I could see the swollen section in the middle of its body, about three feet from the head and four feet from the tail. The python had beautifully reticulated coloring, a fluid pattern of black and brown.

Ahotgiri Camp is an isolated place and the guards spend most of their lives separated from their families. The risk of being attacked by animals or fired upon by poachers is very real and the guns they carry offer limited protection. One of them was an old twelve-gauge with a rusty barrel, the other a .315 rifle like Nasir's, though the sights had broken off. Despite all this Burman and Kakati seemed committed to their work and spoke with pride of the animals they guard, especially the elephants, whom they referred to as "Baba," a term of respect often used for an elder or a saint.

Leaving Ahotgiri Camp, we retraced our route through a stretch of marshland where we had seen nothing that morning. The grass rose up on either side of the road, a corridor twelve feet high, blocking everything from view. Hemanta and I were standing at the back of the jeep, while Nasir and the driver, Krishna, sat in the front. We had our eyes fixed on the road ahead, in case something crossed in front of us. Suddenly I smelled the familiar scent of an elephant's fart. It was unmistakable—a ripe, grassy odor, like rotting mulch.

Krishna seemed puzzled when I tapped his shoulder to stop the jeep, but after he braked and turned the engine off, we could hear a herd of elephants rustling in the grass nearby. They were about fifty meters behind us and off to our left, though we couldn't see them at all. Starting the jeep again, Krishna slowly reversed to a point where we spotted the circular prints of their feet in the dust at the edge of the road. We also saw the discarded leaves of a plant known as tora that one of the elephants had been feeding on and must have dropped.

Waiting for several minutes, we strained to catch sight of the herd but the most we could see was a slight movement in the grass. Nasir quietly worked the bolt on his rifle to load the chamber. The elephants couldn't have been more than fifteen feet away and we heard them snuffling as they fed. At one point, there was the distinct aroma of another fart and this time everyone could smell it. In frustration, Hemanta and I climbed up onto the crossbars of the jeep, as the herd remained tantalizingly close but hidden from view. Finally one of the elephants made a rumbling sound, halfway between a mumble and a groan. From our precarious perch on top of the jeep, we saw the grass begin to shake. A few seconds later, the elephant's trunk appeared, like a periscope rising out of the grass. It curled and swiveled in our direction with its nostrils flared. We had smelled the elephants and now one of them was smelling us. After a minute or two the trunk was retracted and we could hear the herd begin to move away. Obviously they weren't enamored of our scent.

A Book of Elephants

No other part of India is richer in elephant lore than the Brahmaputra valley, and one of the ancient texts of *Gajashastra* comes from Assam, the *Hastividyarnava*. This book shares many similarities with other works like the *Hastyayurveda* and the *Matangalila*. The precise origins of the *Hastividyarnava* remain ambiguous, but the author, Sukumara Barkath ("about whose life or lineage nothing is definitely known"), claims it is based on scriptures originally composed when the world was first created. This seminal text, called the *Gajendracintamani,* is attributed to Princess Vasumati, who lived in a heavenly palace on the slopes of Mount Meru. At the dawn of creation, the princess released a celestial sow into the forests. At the same time, Indra let loose a boar, and the two animals began grazing together. Coming upon a creeper with leaves shaped like elephants' ears, the pair ate these up, after which they mated. Twelve months later the sow produced a litter of elephants, the first of that species.

Years passed, and as the elephants multiplied, they began to raid the palace gardens. After chasing them away on three occasions, Indra transformed two of the herd into white elephants, which he adopted as his mounts. These were the only elephants suitable for a king and a record of their history and traits was posted at the gates of Mount Meru. Centuries later, this sacred knowledge of elephants was transmitted to man through a pious Brahmin named Changjasi, who traveled to Mount Meru and copied down the treatise word for word.

One of the few existing copies of the *Hastividyarnava* is kept on Majuli Island in the center of the Brahmaputra. Ferries cross over to Majuli from Nematighat, a couple of hours' drive northeast of Kaziranga. At this point the river is immense, well over a kilometer in breadth, a sprawling current that divides into three or four channels with sandbars in between. Our van and two jeeps were squeezed onto the deck of an overloaded ferry. The roof of the boat was lined with more than twenty motorcycles and scooters. Every other inch of space was occupied by passengers, and setting off from shore, we looked like a floating traffic jam. Because of the sandbars, which had to be circumnavigated, it took another two hours before we reached Majuli. Along the way, I watched Gangetic dolphins surfacing in the current like humped gray waves. These riverine cetaceans are virtually blind though it hardly seems a disadvantage, for the turgid waters of the Brahmaputra are clouded with silt. Similar in size to a saltwater dolphin, *Plantanista gangetica* averages seven to eight feet, with a long snout and jaws that contain as many as a hundred teeth.

Unlike the barren sandbars, which shift from year to year, Majuli is a permanent island, though it is gradually shrinking because of erosion. Sections are inundated by annual floods, but much of the land is cultivated and there are scattered villages and settlements with a combined population of over 135,000. One of the distinctive aspects of Majuli is the satras, often described as monasteries or spiritual retreats. The first satra was founded by Shankardeb, a Vaishnavite saint who visited the island in the fifteenth century; he is generally referred to as the father of Assamese culture.

At the gate of the Kamalabari Satra, Hemanta and I were greeted by a handsome, bare-chested man wearing only a white cotton dhoti wrapped around his waist. He had smooth, clean-shaven features and his long hair was twisted into a loose knot at the nape of his neck. Every gesture and movement he made was graceful and controlled, with the poise of a dancer. His bare feet hardly seemed to touch the ground as he led us through a grove of trees sheltering the satra. Following him, I felt as if we were stepping back in time, several thousand years, into a mythical Vedic age.

Dulal Saikia explained that he has lived at the satra for thirty-three years, having joined the monastery at the age of seven. All of the monks at the Kamalabari Satra are trained as dancers and actors so that they can take part in the sacred Bhawana, a form of religious theater that celebrates the many incarnations of Vishnu. Saikia and the other monks live an ascetic life, adhering to a code of principles laid down by Shankardeb. Among other things, the men remain celibate and refrain from drinking alcohol and eating meat. At the same time, these monks do not live in isolation, and many of them have jobs outside the satra, working as schoolteachers or government clerks; a few even run shops and businesses, though they are forbidden to farm the land. Like most of the monks, Saikia shares his quarters with an older mentor and a young acolyte of about thirteen. Most of the satras are organized into these fraternal units and follow a tradition of apprenticeship.

In the temple and performance hall Saikia pointed out an image of the elephant. Carved in wood, it is part of the shrine that contains the *Bhagavad Gita*. At the base of the shrine is a tortoise, in the center a tusker, and on the top a lion. Saikia explained that the tortoise symbolizes the earth. The elephant represents "the wild and uncontrollable mind," and the lion stands for "the power of prayer"—nam. Only this lion is able to subdue the elephant, as it represents the discipline and devotion of an ascetic life.

The *Hastividyarnava* is located in the library of another monastary, the Auniati Satra. With the help of Dulal Saikia, Hemanta had arranged for me to view it. We drove across the island on a badly rutted road, through paddy fields and bamboo groves. The outer gate of this satra had two plaster elephants raising their trunks overhead, while the inner gate was guarded by a pair of brightly painted lions.

Once inside, we were taken to a small museum, which contained an odd assortment of objects—several tarnished swords and musical instruments made of brass and copper, as well as an ivory-handled walking stick and a chess set made of ivory. Hanging on one wall was a framed mat woven out of thin strips of ivory shaved off a tusk, each strip cut to a breadth of less than half a centimeter. It was similar to bamboo mats that are made throughout Assam, with intricate patterns emerging out of the woven strands. The workmanship on the ivory mat was exquisitely detailed, though the overall effect was not very different from that of cheap white plastic.

Beside the museum was the library, a single, rectangular room with bookcases on all four sides. These contained ancient as well as modern copies of the *Bhagavad Gita, Bhagavad Purana, Ramayana,* and *Mahabharata.* On a table in the center of the library lay a wooden chest with floral carvings on its lid. The caretaker, who accompanied us, opened it carefully and folded back the strips of silk in which the book was wrapped. Though I knew what to expect, I was still startled by the vivid colors on the first page, a miniature painting of an elephant carrying a king. Turning over each leaf of the book, we saw pictures of different elephants, some of them wild, others being trained. On one side of each page was a block of text written in Assamese. There were a total of fifty-five pages in the box, each of them about twelve inches long and six inches wide. These are made from the bast, or inner bark, of the agarwood tree. The bark is flattened and specially treated to provide a smooth, rigid surface. Each page is about as thick as a piece of light cardboard. Agarwood, known as sasi in Assamese, is extremely valuable and is used as incense.

Another copy of the *Hastividyarnava* is housed in the library of Assam's Department of Historical and Antiquarian Studies at

Guwahati. It has been translated into English by Pratap Chandra Choudhury, who believes it was commissioned in 1734 by the Ahom king, Siva Sinha, and his queen, Ambika Devi. The *Hastividyarnava*, like other classical Indian texts, was written for the erudition and entertainment of kings and courtiers, but it includes elements of folklore that come out of the indigenous traditions of tribal hunters and mahouts. Much of the *Hastividyarnava* is a series of lists giving the different characteristics of elephants, particularly those that are suitable for a king. The book also includes instructions on how the elephants are to be handled, particularly where they should be "pierced" with a goad or ankush. These details reflect an elaborate system of pressure points that are used to either subdue an elephant or drive it into a frenzy so that it will be more effective in battle. Toward the end of the book are unillustrated pages devoted to medicinal herbs and other remedies used for treating elephants when they fall sick. Some of these are derived from Ayurvedic traditions, while others owe more to Tantric rituals, accompanied by magical incantations.

Though I had previously read Choudhury's translation of the *Hastividyarnava* and seen color reproductions of the paintings, there was something very different about holding the original text in my hands. The agarwood pages were almost weightless, yet I was hesitant to touch them. The imagery of each painting had an iconic brilliance, like framed moments in a dream. I could not read the script but the story was familiar. There were pictures of elephants in the wild, a herd with calves near a water hole. Other paintings depicted hunters stealing through the forest and elephants tied with ropes. Mahouts carried goads in their hands as they trained the animals in violent acts of war. Kings and queens sat in ornate howdahs, carried through the streets on the backs of royal tuskers. Seeing the *Hastividyarnava* at the Auniati Satra, I understood more than ever before how the lore of the elephant in India is a testament to this species, a tribute paid by human beings to a creature far greater than ourselves. The mythology of elephants, as recounted by humans, may be full of allegory and other poetic devices, but in essence it is a legend that comes from nature itself, like the bark on which it is written and the vegetable dyes that give it color.

Lingering on the pages of this book was the faintest odor of incense, though the original resins in the agarwood must have faded long ago. It was a sweet, musty perfume that may have come from the cloth in which the book was wrapped, though I imagined another source. The *Hastividyarnava* contains a passage that describes the finest elephants of all, which, Sukumara Barkhat tells us, "make the forests they dwell in fragrant. These elephants are very fine to look at. Other elephants get pleasure in coming into contact with them. The entire forest remains in its fullness because of such elephants."

Rescue and Release

Not far from the entrance to the eastern range of Kaziranga is an animal rescue center run by the Wildlife Trust of India. I had been told about the center when I visited WTI's headquarters in Delhi and I was eager to see it, particularly after having observed the two elephant calves that were being reared at Chilla. However, when I reached the rescue center, there was a large sign:

No Visitors

As I hesitated near the gate, a man emerged from one of the buildings inside and looked at me with suspicion. Introducing myself, I explained that I wasn't interested in photographing any of the animals and only wanted to speak with the project veterinarian, Dr. Choudhury, whose name I had been given.

"I am Bhaskar Choudhury," he said. "Please wait a minute."

Through the gate I caught a glimpse of a baby rhino and a

buffalo calf, both of whom were watching us with curiosity. Before letting me inside, Choudhury gave instructions to one of his staff, who shooed the two animals away behind a canvas blind.

"I'm sorry about that," he said. "But you understand, don't you? We try to shield the animals from too much exposure to human beings."

Baby animals invariably attract attention and with the number of tourists visiting Kaziranga, the rescue center could easily become a petting zoo. This would jeopardize any possibility of reintroducing these animals into the wild, which is why Choudhury and his staff keep their gates firmly closed.

Though the project began in 2000, construction of the center had just been completed. Along with the animal enclosures, there is a surgery room and offices. WTI is a nongovernmental organization, which operates the center in collaboration with the Assam Forest Department. The center is funded by the Government of India's Animal Welfare Department and the International Fund for Animal Welfare.

When I asked Dr. Choudhury how the relationship worked, he said, "The forest department is quite happy to hand the animals over to us because it relieves them of the responsibility and expectations of the public." The animals remain legally under the authority of the forest department but are given care and ultimately released by the staff of the rescue center.

Located near one of the corridors between Kaziranga and Panbari Forest Reserve, the center is well situated to deal with animals displaced by floods, as well as those that are injured in collisions with vehicles along the highway.

"One of the first things we do when an animal is rescued is try to reduce the amount of stress it experiences during capture," said Choudhury. "We cover their eyes and ears as best we can. Often we give them sedatives. After that we provide first aid as necessary and try to return the animal to the wild as soon as possible."

The center is committed to respond to any call for wildlife rescue in the Northeast, though distances and communication are a challenge, particularly in places like Arunachal Pradesh. Never knowing what crisis to expect, Choudhury and his staff often have to improvise solutions as they deal with everything from hog deer hit by trucks to immature adjutant storks that have fallen out of their nests. Most of their work occurs during the monsoon, when floods drive the animals out of their usual habitat. The worst floods in recent memory occurred in 1988, when a total of sixty rhinos died at Kaziranga, along with hundreds of other animals.

In August 2000, soon after the project started, Choudhury had his first experience rescuing a baby elephant.

"A report came in that a calf had been separated from a herd and was stranded in a patch of water hyacinths. It was only about a month old and severely dehydrated by the time we got there. The umbilical cord had also become infected. With the help of forest department staff, we gave it intravenous fluids. We also treated the infection with antibiotics. At the same time a herd was located nearby and using two tame elephants, we guided the calf toward them. To mask any human smells we smeared the calf with mud and elephant dung. When it was about a hundred meters from the herd, the calf called out and one of the females came forward, while the tame elephants moved back. The female was lactating and the baby immediately began to suckle. Others surrounded them and sniffed at the calf. Eventually, it went off with the herd."

Despite this apparent success, Choudhury is cautious about drawing conclusions.

"I can't say for sure that this was the calf's mother or that it even survived, but we did get reports afterwards that the herd had three calves and it is likely that one of these was the animal we rescued. After that experience we have not had such good luck. Out of the nine elephant calves rescued since then, only two have survived. Both are being raised here at the center."

The ideal time to release a rescued calf is immediately after it has been given first aid, before it has had prolonged contact with human beings. In many cases, though, the animals are too badly injured, weak, or sick to rejoin a herd. They have to be brought back to the rescue center, after which release becomes less possible the longer they are kept in captivity.

Providing medical treatment for wild animals can often be a challenge. For instance, an elephant is given an intravenous drip through a vein in the back of its ear. Intravenous feeding is not always possible, however. The rhino calf, which I saw briefly on arriving at the center, gave Choudhury a lot of trouble when it was rescued because it had pneumonia. He couldn't put it on a drip for fear that this would add to the fluid in its lungs. The rhino had to be nursed by hand and he said they were lucky it survived. As we were talking, Choudhury showed me photographs of other animals he has treated at the center, including one of the riding elephants from the park, who had a gash in his trunk.

"It was very difficult because the elephant was in a lot of pain and wouldn't let us touch the wound. It's also impossible to put sutures in an elephant's trunk because it requires so much movement and that's the only way it can drink. Fortunately, with the help of antibiotics, the wound has healed and the elephant is all right."

A chart on the wall of Choudhury's office listed the animals housed at the center. These included the two baby elephants, the rhino and buffalo calf, as well as a barking deer and a pig-tailed macaque monkey. One of the first things I noticed was that none of the animals had been given names. They were identified only by their Latin names as well as their age. When I mentioned this to Choudhury, he nodded and said that this was part of the philosophy of the rescue center. In their interaction with the animals he and his staff avoid the impulse to treat them as pets. Being a scientist and veterinarian, Choudhury is committed to the welfare and survival of the animals he treats, but he is equally aware that human beings tend to anthropomorphize the needs of other creatures. While acknowledging the altruistic and emotional response we have for animals, especially those that are injured or at risk, he cautions: "We must not interpret their behavior according to our own."

The Rhinoceros and the Buffalo

Among the first visitors to Kaziranga, after it was declared a sanctuary in 1938, was E. P. Gee, the manager of a nearby tea estate. In his book, *The Wild Life of India,* Gee describes the place as "one of the last unspoilt and unoccupied grassy areas of the Brahmaputra valley . . . a sort of *terra incognita.*" He also notes that much of the forested land in Assam had been cleared and occupied by tea gardens like his. When Gee first tried to get permission to visit Kaziranga he was told, "No one can enter the place. It is all swamps and leeches and even elephants cannot go there." In fact, these reserve forests were freely accessible to poachers, who operated in the area with impunity.

While most tea planters occupied their leisure hours in pursuit of shikar, E. P. Gee preferred the camera to the rifle and spent much of his free time taking photographs of animals in Kaziranga. As an active member of the Bombay Natural History Society and the Indian Board for Wildlife, he helped promote awareness for nature conservation in India. When Jawaharlal Nehru and Indira Gandhi visited Kaziranga in 1956, E. P. Gee was there to accompany them on a tour of the park. He took a picture of Nehru riding through the grasslands on an elephant, an image that projects one of the many priorities of India's first prime minister. In his foreword to Gee's book, Nehru wrote:

Wildlife? That is how we refer to the magnificent animals of our jungles and to the beautiful birds that brighten our lives. I wonder sometimes what these animals and birds think of man and how they would describe him if they had the capacity to do so. I rather doubt if their description would be very complimentary to man. In spite of our culture and civilisation, in many ways man continues to be not only wild but more dangerous than any of the so-called wild animals. . . .

I cannot say that we should preserve that form of wildlife which is a danger in our civilized haunts or which destroys our crops. But life would become very dull and colourless if we did not have these magnificent animals and birds. . . . Our forests are essential for us from many points of view. Let us preserve them. As it is, we have destroyed them far too much.

On a bright, clear morning, very similar to the day forty-seven years earlier when Nehru came to Kaziranga, I watched a crowd of visitors gather at dawn for a wildlife tour on elephant back. Most of these tourists were Assamese, though there was also a group of students from neighboring Nagaland. About twenty adult elephants were lined up near the mounting station, along with three calves, one of them suckling at its mother's breast. Curling up its trunk, the calf burrowed in between the cow's forelegs to find her teats. The rest of the elephants stood by patiently while each was loaded up in turn. One of them was a large bull, whose trunk was draped nonchalantly over his left tusk. In contrast to the subdued elephants, the crowd of tourists was full of excitement—shrieks of laughter, voices raised, children darting about, cameras flashing. Once these visitors had climbed onto the elephants, however, they grew immediately calm and silent. As the tame herd set off in single file, the only sound I could hear was the swish of grass against their legs.

Spreading out in loose formation they moved across an open plain, past herds of grazing swamp deer and a few scattered buffalo and rhino. Though these elephant rides are restricted to a limited area near the gate of the park and the tours last only an hour, there is still something grand about the spectacle. Twenty elephants walking in a line is a dramatic sight, especially with the panorama of the Brahmaputra valley as a backdrop. At sunrise, the distant foothills and white snow peaks of the eastern Himalayas stood out against a polished sky, one range of mountains leading on to the next. It would seem impossible for anyone to witness this sight and fail to understand the significance and value of India's natural heritage.

That same day I climbed a watchtower overlooking Dunga Lake and counted twenty rhinoceros grazing along the shore. At least two dozen buffalo were also scattered across the plain. For half an hour or so, the animals barely stirred, like chess pieces positioned on a board waiting for the next move. The stillness of the scene was only broken by the occasional splash of a fish feeding on the surface of the lake. I could have taken a photograph when I first arrived in the watchtower, then another as I left, and it would have been difficult to find any difference between the two. The light may have changed slightly and the shadows shifted but the animals remained in place, as if the landscape lay in perfect balance. It is a scene that would have been common three centuries ago, from one end of the Himalayan foothills to the other, from the Indus to the Brahmaputra. Today the only place in the subcontinent where wild buffalo and rhino graze side by side is Kaziranga and a few other pockets of Assam.

The great Indian one-horned rhinoceros has been persecuted even more than the tiger or the elephant. Babur, the first Mughal emperor, records hunting these animals soon after he crossed over the mountains from Afghanistan. Until the end of the eighteenth century, rhinos were shot and killed throughout most of northern India until they were virtually exterminated. Unlike the elephant, the rhinoceros serves no useful purpose in captivity; neither does it have any great religious significance. For many, the only value of this animal lies in its horn, "a closely-matted mass of horny fibre issuing from the skin," as S. H. Prater describes it in *The Book of Indian Animals*. The persistent belief that rhino horns have medicinal properties continues to drive an illegal trade in India and other parts of Asia.

Poaching and the spread of agriculture are not the only ways in which human beings have put the rhino and its habitat at risk. The grasslands and marshes of Kaziranga are threatened by invasive weeds like the water hyacinth, imported from America, which now chokes many of the ponds and lakes in the park. Equally insidious is a thorny kind of ground cover, *Mimosa invisa*, introduced into Assam by tea planters because of the plant's ability to fix nitrogen in the soil. Spreading beyond the tea gardens, it is now replacing indigenous grasses in the park on which rhinos, elephants, and other animals depend for food.

At one of the forest camps in Kaziranga, I saw the skull of a rhinoceros that had died of natural causes. The horn had been removed, but there was a bump at the end of its snout where this would have rested. Contrary to most people's assumptions, the rhino seldom uses its horn for fighting. Instead it is equipped with large tushes, somewhat like a wild boar's, which can inflict a fatal wound. Most of the time, though, rhinos live a peaceful, if misanthropic, existence, feeding almost constantly on grass and waterweeds.

Driving through the western range of Kaziranga, we came upon one rhino about thirty meters from the road. He stood there with a look of absolute boredom, as if having seen too many tourists that day. His movements were slow and ponderous, unlike those of the white egret that fidgeted on his back. The rhino seemed to be weighed down by his heavy armor plating, the folded gray layers of his hide "studded with masses of rounded tubercles," as Prater puts it, which look like rivets on a tank. There is something undeniably prehistoric about the rhinoceros that reminds us of its predecessors, *Rhinoceros sivalensis* and *Rhinoceros paloeindicus,* which used to roam the same belt of grasslands girdling the Himalayas.

Compared to rhinos, wild buffalo have suffered a very different fate. Because this species was easily domesticated and provided milk, *Bubalus bubalis* has virtually disappeared in the wild while becoming one of the most prevalent varieties of cattle in India. Unlike the elephant, buffaloes have no reluctance to procreate in captivity, and through selective breeding they have been transformed into docile and sedentary beasts, unlike their more aggressive relatives in the wild. Over many generations the spreading horns of the buffalo have been reduced to more compact headgear, and most tame animals have a bulky, awkward appearance. Wild bulls at Kaziranga occasionally mate with village buffaloes along the periphery of the park and the calves are generally indistinguishable.

One of the many buffalo that I saw at Kaziranga was a large bull standing at the edge of a muddy pond. His horns stretched out three feet on either side of his head, curving up at the ends like scimitars. Hanging from the tip of one of these horns was a loop of water hyacinths that he had snagged while drinking from the pond. The leaves and purple flowers looked like a soggy corsage. Glaring at us suspiciously, the buffalo flared his nostrils. His hide was black and glossy, as if recently buffed with boot polish, and his muscles flexed beneath his skin. After watching us for awhile, the buffalo took a few steps forward and then lay down. Rolling onto his back and wallowing with his hooves in the air, he covered himself completely in mud. When he stood up, the bull was a dripping gray color, like unfired clay. Glancing contemptuously over his shoulder at us, he set off at a trot.

Watching this bull, I was reminded of the herds of buffalo I had seen in Rajaji National Park, tended by Gujjar herdsmen. Here was the same animal—but while the buffaloes at Rajaji represent a severe imbalance in the environment, encroaching on the habitat of elephants and other animals, at Kaziranga they are an integral part of the landscape. As man's relationship with the rhino and the buffalo demonstrates, our impact on wildlife varies drastically from species to species. On one hand we have killed off all but a handful of rhinos that now face extinction, while on the other our success at taming and breeding buffaloes has increased the population in captivity but severely reduced numbers in the wild. Human beings are responsible for the fate of both species—for the rhino it means decimation, for the buffalo domestication.

The future of the elephant may lie somewhere in between, but it is important to remember that none of these species, including Homo sapiens, exists in isolation and to survive they must be able to coexist. As I look back on the scene I witnessed from the watch-tower at Dunga Lake, the image of a chessboard remains fixed in my mind. The rhinos and buffaloes were poised like the pieces in an unfinished game, arrayed seemingly at random but actually positioned according to complex patterns of nature. The stillness of the landscape revealed a tension that held everything in place, as if waiting for human beings to make the next move.

L'Envoi

On our last day at Kaziranga we explored the western range of the park. It was late afternoon and fading sunlight had turned the grasslands a coppery green. The Karbi Hills were folding into shadows and a flight of bar-headed geese unraveled in loose formation against a burnished sky. Ahead of us, a rhino came blundering up the embankment and onto the forest road. He turned to glare at the jeep before descending into the grass again.

My travels in pursuit of elephants were almost over and it was difficult not to feel regret. The memories and stories would always stay with me but not the immediate sensations of the forest and those unexpected moments of discovery. A year earlier, on my final day at Corbett Park, I had almost given up on finding any elephants; then there they were, on the ridge in front of me. Over the past year I had seen elephants in the wild and in captivity, in forest camps, zoos, and temples. I had also seen countless images of these animals painted on the walls of caves, on paper, cloth, and bark, or carved out of ivory, wood, and stone, as well as cinematic visions flickering across a video screen. On the rock at Kalsi my fingers had traced the outline of Gajatame and in Mumbai I had watched Ganesha dissolve into the sea.

But as we drove through the grasslands of Kaziranga, all of that was behind me and my eyes were focused only on the forest ahead. The sun was now setting beyond the hills as we turned off the main road onto a narrow, overgrown track. Less than a minute later a herd of elephants appeared in a clearing at the edge of the trees. They were moving through the grass at a deliberate, unhurried pace, while feeding silently. Like probing fingers, their trunks sought out the tender basal shoots and uprooted them with selective ease. Ears flapping, they glanced at us from time to time but seemed untroubled by our presence. The elephants were of all ages, from calves no more than a few months old, to juveniles and young adults, as well as many full-grown cows. Altogether there must have been thirty animals in the herd.

Though the distance between us was less than fifty meters, the elephants seemed much farther away, as if separated in time and moving through a different world, a different kind of history than ours. They seemed older than any myths or legends, detached from any symbols we might assign—gray shapes with supple trunks and waving ears that lay beyond the reach of human voices, communicating among themselves in low, inaudible tones. Their memories were not the same as mine, but more tangible and tactile, recalling the familiar scent of mud by a lakeshore, the succulent leaves of a ficus tree, the sweet pith inside the stem of a tora plant. Instinctual. Inhuman.

One of the elephants was pregnant, her flanks swelling out on either side, so that facing us she looked twice as large as the others. This cow must have been ready to give birth and her movements seemed restrained, slowed down by the extra weight she carried. Even with the jeep's engine turned off, we could hear no sound from the herd, except for the occasional whisper of grass. For twenty minutes or so, we watched the elephants in silence, as darkness settled over the Brahmaputra valley and muted colors blurred into gray. By this time the last stragglers in the herd had vanished into the forest as the night began its own gestation.

acknowledgments

Elephants express gratitude by flapping their ears and waving their trunks appreciatively. Being ill-equipped to offer my thanks in such a manner, I must rely on inadequate gestures. I am extremely grateful to the John Simon Guggenheim Memorial Foundation for a fellowship that permitted me to complete work on this book. Many friends, acquaintances, and strangers have offered assistance, advice, and information as I traveled to different parts of India and spent time researching the elephant's story in libraries and museums. I am especially indebted to the following individuals for their help, though none of them should be held responsible for any errors or shortcomings in the book: Zahur Ahmed, Shalini Agarwal, Carol Evans and Tom Alter, N. V. K. Ashraf, Parbati Barua, André Bernard, S. S. Bist, Jeffrey Campbell, S. Chandola, Kamla Chowdhry, John Coapman, J. C. Daniel, Hemanta Das, Ranjit Das, David Davidar, Priya Davidar, Jill Grinberg, David Hough, Vineeta Jalan, Maya Jhangiani, A. J. T. Johnsingh, V. K. Karthika, Monica Law and Kevin Mansfield, Kadambari Mainkar, Sanjaya and Ajay Mark, Vivek Menon, Andrea Schulz, Brijraj Singh, Devika and Kavi Singh, Lillian Singh, Sushil Singh, Vivek Sinha, Woodman Taylor, Thomas Trautmann, Saroja and Govindan Unny, John Wakefield. My wife, Ameeta, and our children, Jayant and Shibani, have been especially supportive, tolerating my absences from home and putting up with more elephant stories than anyone would ever want to hear.

notes

PAGE

VII: *"Why no decorated ... a tranquil journey?"* Ramayana, trans. E. Gerow in *The Literatures of India: An Introduction,* E. C. Dimock et al. (Chicago: University of Chicago Press, 1974), 59.

VII: *"Few things provoke ... of ancestry."* Barry Lopez, *Arctic Dreams* (New York: Bantam, 1986), 37.

prologue
PAGE

2: *The fourteenth century ... able to speak.* Muhammad al-Damiri, *Hayat al-Hayawan,* trans. A. S. G. Jayakar (Bombay: D. B. Taraporevala, 1906), 568.

3: *"who is gentle ... of perfect sensitivity."* Nilakantha, *Matangalila,* trans. Franklin Edgerton (New Haven: Yale University Press, 1931), 79.

I
winter sanctuary
PAGE

11: *Newspaper articles ... bled to death.* Poornima Joshi, "Slow, Painful Death at Corbett," *Hindustan Times,* 21 January 2001.

11: *Later investigations ... is common.* A. J. T. Johnsingh and A. Christy Williams, "Bull Elephants in the Rajaji-Corbett Range," *Sanctuary* (June 2002): 45.

12: *Within a few months . . . poachers.* "Poachers Kill Ranger, Injure Three Guards in Corbett," *Times of India,* 30 August 2001, 7.

15: *"elephants were . . . not shot at."* D. K. Lahiri-Choudhury, *The Great Indian Elephant Book* (Delhi: OUP, 1999), xxvi.

16: *"More than 'preservation' . . . vital natural resource."* Lahiri-Choudhury, *The Great Indian Elephant Book,* xxiv.

16: *"A tiger is . . . her fauna."* Jim Corbett, "On Man-eating," in *Jim Corbett's India,* ed. R. E. Hawkins (Delhi: OUP, 1978), 247.

17: *"The reactions . . . of a peregrine falcon."* Jim Corbett, *The Temple Tiger* (London: OUP, 1955), 128.

21–22: *Though an elephant . . . silently as it does.* Martin Saller and Karl Groning, *Elephants: A Cultural and Natural History* (Cologne: Konemann, 1999), 69.

26: *A number of instances . . . delivered a calf.* E. P. Gee, *The Wild Life of India* (London: Fontana, 1969), 194.

II
on the origin of elephants

PAGE

34: *The poem opens . . . his beloved yakshi.* Kalidasa, *The Meghduta of Kalidasa,* trans. M. R. Kale (Delhi: Motilal Banarasidass, 1991), 1–8.

34–35: *"Exiled in the forest . . . of her embrace."* Kalidasa. This rendition is based on English translations and commentary in several texts: *The Meghduta,* Kale, 1–8; *The Megha-Duta of Kalidasa,* Sushil Kumar De (Delhi: Sahitya Akademi, 1957), xxix–xxx; *The Meghaduta of Kalidasa,* G. H. Rooke (London: OUP, 1935), 2–3; and *The Megha Duta or Cloud Messenger,* H. H. Wilson trans. (London: Black, Parry, and Co., 1814), 23–24.

35: *For much of . . . elephant's hide.* Kalidasa, *The Meghduta,* Kale, 64.

35: *Though musth can occur . . . in search of a mate.* Raman Sukumar, *Elephant Days and Nights* (Delhi: OUP, 1994), 37–40.

36: *battles an elephant demon. . . . Gajasura's skin.* Kalidasa, *The Meghduta,* 69–70.

36: *The metaphor . . . an image of fertility.* H. Zimmer, *Myths and Sym-*

bols in Indian Art and Civilization (Princeton: Princeton University Press, 1974), 108–9.

36: *"a laden cloud . . . flashes gleam."* *The Panchatantra,* trans. A. W. Ryder (Chicago: University of Chicago Press, 1926), 311.

37: *"A peculiarity noticed . . . from the temporal glands."* M. Krishnan, "Musth" in *The Asian Elephant,* J. C. Daniel (Bombay: BNHS, 1998), 127–28.

37: *In the* Ramayana *epic . . . juices of royal elephants.* R. K. Narayan, *The Ramayana* (New York: Penguin, 1972), 23.

38: *"With honey-colored nails . . . at a distance."* Nilakantha, *Mantangalila,* 82–83.

38: *"a cloud that . . . of water."* Nilakantha, *Mantangalila,* 85.

38–41: *When the cosmic egg was . . . "and bad for them,"* Nilakantha, *Mantangalila,* 41–46.

41: *Palakapya provides . . . "brooding melancholy."* Nilakantha, *Mantangalila,* 49.

41–42: *Palakapya also divides . . . like a deer.* Nilakantha, *Mantangalila,* 48–49.

42: *In one particular . . . dung and urine.* Nilakantha, *Mantangalila,* 53.

42: *Aeons later . . . one generation to the next . . .* Raman Sukumar, *Elephant Days and Nights: Ten Years with the Indian Elephant* (Delhi: Oxford, 1994), 55–56.

42–44: *Out of dozens . . . put into practice.* Martin Saller, *Elephants,* 58–59 and 308–9; Robert Delort, *The Life and Lore of the Elephant* (London: Thames and Hudson, 1992), 20–26.

44: *For many years . . . question of survival.* Saller, *Elephants,* 431; Richard Sabin, London Museum of Natural History, correspondence with author.

46: *In the Andaman archipelago . . . on land.* Saller, *Elephants,* 18–19; A. P. Gaeth et al., "The developing renal, reproductive, and respiratory systems of the African elephant suggest an aquatic ancestry." *Journal of Developmental Biology* 96, no. 10; available at http://www.pnas.org/cgi.

47: *a first-century B.C. sandstone . . . head of an elephant.* K. S. Srivastava,

The Elephant in Early Indian Art (Varanasi: Sangeeta Prakashan, 1989), fig. 73.

47: *"could go anywhere . . . any shape."* Nilakantha, *Mantangalila,* 44.

48: *Like most elephants . . . one-eyed giant.* Saller, *Elephants,* 36.

48: *nature provides elephants . . . starves to death.* S. H. Prater, *The Book of Indian Animals* (Bombay: BNHS, 1948), 223; Sukumar, *Elephant Days and Nights* (1994), 107.

49: *approximately 40 percent . . . about 10 percent.* R. Sukumar, *The Asian Elephant: Ecology and Management* (Cambridge: Cambridge University Press, 1989), 165–66.

49: *Colonial sportsmen . . . shooting a makhna as well.* A. J. W. Milroy, *Management of Elephants in Captivity,* ed. S. S. Bist (Dehradun: Natraj, 2002), 107–13.

50: *"If tuskers . . . in its tusks."* Sukumar (1994), 145.

50: *There is also the possibility . . . responsible for tusks.* Vivek Menon, *Tusker* (Delhi: Penguin, 2002), 114.

51: *to illustrate . . . varying climates.* Charles Darwin, *The Origin of Species* (New York: Mentor, 1958), 138.

51: *"The elephant is reckoned . . . the first pair."* Darwin, *Origin,* 138.

51: *"The Indian elephant is known sometimes to weep."* Charles Darwin, *The Expression of the Emotions in Man and Animals* (London: William Pickering, 1989), 126.

51–52: *"tears which suffused . . . down his cheeks."* Darwin, *Expression of Emotions,* 126.

52: *In much of the chapter . . . did not contract.* Darwin, *Expression of Emotions,* 127.

52–53: *The most beautiful . . . sometimes fall out.* Nilakantha, *Mantangalila,* 92–104.

53: *During this conflict . . . with its tusk.* Narayan, *Ramayana,* 160.

53: *Much of the tragedy . . . for an elephant.* C. Rajagopalachari, *Ramayana* (Bombay: Bharatiya Vidya Bhavan, 1996), 94–99.

55: *Confronted for the first time . . . like trombones.* Saller, *Elephants,* 240.

55: *the Sanskrit . . . the monsoon.* Monier Monier-Williams, *A Sanskrit-English Dictionary* (Delhi: Asian Educational Services, 1999), 342.

55: *The modern Hindi... versatile trunk.* Monier-Williams, *Sanskrit-English,* 1294–95.

56–57: *"The largest land animal... sun and moon." "They themselves... in the ground." "tremble in fear of an ambush." "Why should... succumbed to luxury."* Pliny, *Naturalis Historia,* book 8, trans. H. Rakham (Cambridge: Harvard UP, 1940), 3–29.

57: *During the second century... rendered helpless. Physiologus,* trans. Michael J. Curley (Austin: University of Texas Press, 1979), 29–32.

57: *Twice round an elephant's foot... its height.* R. S. Morris, "Size," *Journal of the Bombay Natural History Society* 24 (1920): 800, quoted in *The Asian Elephant,* J. C. Daniel (Dehradun: Natraj, 1998), 29.

58–59: *Early British efforts... fourteen inches.* L. C. Miall and F. Greenwood, *Anatomy of the Indian Elephant* (London: Macmillan, 1878), 61–66.

59: *"Al-Kazwini states... nothing is impossible!"* al-Damiri, *Hayat al-Hayawan,* 569.

59–60: *"In copulation... appears in sight."* Samuel Johnson, *Dictionary of the English Language,* cited in *The Professor and the Madman,* Simon Winchester (New York: Perennial, 1999), 90.

60: *"informs us... standing posture."* Miall and Greenwood, *Anatomy,* 66.

60: *"the fully erect penis"... a couple of hours.* Sukumar, *Elephant Days and Nights* (1994), 42–44.

60: *Sukumar also... with their trunks.* Sukumar (1994), 40–41.

61: *"A transection... of the proboscis."* D. Mariappa, *Anatomy and Histology of the Indian Elephant* (Oak Park: Indira Publishing House, 1986), 188.

61: *"The special type... Ayer's nerve endings."* Mariappa, *Anatomy and Histology,* 189.

III

m y s o r e

PAGE

67: *Across the river... wounded or killed.* N. Baskaran et al., "Home Range of Elephants in the Nilgiri Biosphere Reserve," 296–313 and "Crop Raiding by Asian Elephants in the Nilgiri Biosphere Reserve," 350–67, *A Week With Elephants* (Mumbai: Bombay Natural History Society, 1995).

68: *"Debarking," as it is . . . minerals and other nutrients.* Sukumar, *The Asian Elephant,* 83.

69: *More than any . . .* Jungle Book. Rudyard Kipling, "Toomai of the Elephants," *The Jungle Book* (New York: The Century Company, 1899), 248–53.

69: *"Let us believe . . . Highlander's hoch!"* John Lockwood Kipling, *Beast and Man in India* (London: Macmillan, 1921), 224–25.

75–76: *In 1864 . . . dominated most of his energies.* G. P. Sanderson, *Thirteen Years Among the Wild Beasts of India* (London: W. H. Allen, 1879), 1–4.

76: *"I knew nothing . . . with their habits."* Sanderson, *Thirteen Years,* 101.

77: *"I was determined . . . happened to meet with!"* Sanderson, *Thirteen Years,* 103.

78: *"Elephants are kept . . . please his vanity."* Sanderson, *Thirteen Years,* 88.

79: *Khedah operations . . . "distinguished State guests."* C. Hayavadana Rao, *Mysore Gazetteer,* 1 (Bangalore: Government Press, 1927), 126–30.

79–80: *On November 25 . . . "plied them with."* M. Shama Rao, *Modern Mysore* (Bangalore: Higginbothams, 1936), 147–48.

80: *The last of these . . . volley of gunfire.* Sukumar, *Elephant Days and Nights,* 19–20.

81: *Glossed over by . . . look after the elephants.* Frances Flaherty, *Elephant Dance* (New York: Scribners, 1937), 59.

83: *Though Sabu . . . with him by ship . . .* "Forgotten Hollywood Star from India," *The Hindu* (18 February 1990, Bangalore ed.), 20.

IV

remover of obstacles

PAGE

86–87: *Dating back to . . . entrance to the temple.* Myriam Kaye, *An Illustrated Guide to Bombay and Goa* (Hong Kong: The Guidebook Company, 1990), 99.

89: *(The Portuguese . . . target practice.)* Kaye, *Illustrated Guide,* 99.

90: *Every schoolchild . . . British authorities.* Neeta Kolhatkar, "Tilak Revisits Girgaum!" *Times of India,* 30 August 2001, special supplement, 1.

91–92: *Many of the mandals . . . within the community.* Neeta Kolhatkar, "Kal Aaj Aur Kal," *Times of India,* 29 August 2001, special supplement, 1.

95–96: *Several different . . . was an elephant.* Robert L. Brown, *Ganesh: Studies of an Asian God* (Delhi: Sri Satguru, 1997), 2–3 and Shakunthala Jaganathan and Nanditha Krishna, *Ganesha* (Mumbai: Vakils, 2000), 11–17.

96: *Ganesha is often . . . sound of om.* Jaganathan and Krishna, *Ganesha,* 93.

97: *Most images of Ganesha . . . girdling his midriff.* Jaganathan and Krishna, *Ganesha,* 19–23.

97: *According to one myth . . . in his honor.* Jaganathan and Krishna, *Ganesha,* 20.

98: *During the Ganapati festival . . . 1,200 rupees per kilo.* Kishore Rathod, "51 Ways Modaks Can Delight," *Sunday Mid-Day,* 26 August 2001, 5.

99: *Archeologist . . . remover of obstacles.* M. K. Dhavalikar, "Ganesa: Myth and Reality," in *Ganesh: Studies of an Asian God,* ed. R. Brown (Delhi: Sri Satguru, 1997), 57.

99–100: *"The most striking . . . merger takes place."* A. K. Narain. "Ganesa: The Idea and The Icon," in *Ganesh: Studies of an Asian God,* ed. R. Brown (Delhi: Sri Satguru, 1997), 31.

101–2: *In a memoir . . . become an ornithologist.* Salim Ali, *The Fall of a Sparrow* (Delhi: Oxford UP, 1985), 7.

103–4: *"Conserving the elephant means conserving the human environment."* J. C. Daniel, "Conservation and Future of the Indian Elephant" (Inaugural Address, National Seminar and Exhibition, "The Call of the Elephant," Kolkata, 18–19 August 2001).

104: *"We were . . . thoughts were bent."* Ali, *Fall of a Sparrow,* 225–26.

105–6: *During the festival . . . gestures of unity.* R. Sridhar, "Make Way for the Lalbaughcha Raja," *Times of India,* 30 August 2001, special supplement, 3.

V

queſtionſ of captivity

PAGE

115: *scientists have . . . on which they feed.* Sukumar, *The Asian Elephant,* 69.

115: *Ancient precedents . . . could not be ignored.* E. S. Varadarajaiyer, *The Elephant in the Tamil Land* (Annamalainagar: Annamalai University Press, 1945), 66.

119: *"the messenger to the world . . . the world's elephants."* Eric Scigliano, *Love, War, and Circuses* (Boston: Houghton Mifflin, 2002), 251–52.

119: *Loki is described . . . cause earthquakes.* "Loki"; available at http://www.islandia.is/~oldnorse/gods/loki.htm; "Loki," Encyclopedia Mythica; available at http://www.pantheon.org/articles/l/loki.html.

119: *The Tamil Nadu . . . inflicted by farmers.* "Allegations Regarding Cruel Treatment of an Elephant," press release, Embassy of India, Washington, D.C., 17 March 1999.

120: *Maneka Gandhi . . . sensationalized the situation.* Letter from Maneka Gandhi reproduced on IPAN Web site; available at http://www.gcci.org/ipan/loki/maneka_ghandi090799.html (sic).

120: *"He has been . . . compassion for elephants."* Letter from Raman Sukumar, reproduced on IPAN Web site; available at http://www.gcci.org/ipan/loki/raman_sukumar042299.html.

121: *"ecocide" . . . "a joke."* Michael Fox, "New Problems, May 2002"; available at http://www.gcci.org/ipan/loki/newproblems0502.html.

123: *For the past twenty-five years . . . "India's most wanted man."* Sunaad Raghuram, *Veerappan: The Untold Story* (Delhi: Penguin, 2001), xiii.

123–24: *One of the . . . wholesale catalogue.* Raghuram, *Veerappan,* 216–18.

124: *"The lofty hill . . . fiery-eyed elephant."* Varadarajaiyer, *Elephant in the Tamil Land,* 80.

126: *segregated because . . . bring bad luck.* Menon, *Tusker,* 212–13.

128: *sometimes compared to the fragrance of a bride's hair.* Varadarajaiyer, *Elephant in the Tamil Land,* 96.

129–30: *The danger posed . . . and took revenge.* Jacob George, "Trunk Call," *India Today,* 24 February 2003, 42.

132: *Though Kesavan . . . "were well known,";* "Guruvayur Elephants" (2001) available at http://www.guruvayurdevaswom.com.

134: *Periyar has a population . . . all of India.* P. M. Chandran, "Population Dynamics of Elephants in Periyar Tiger Preserve," cited in Menon, *Tusker,* 109.

135: *One of the smallest . . . tied to the forest.* Zacharias P. Thundy, *South Indian Folktales of Kadar* (Meerut: Archana, 1983), ix–xiii.

135–36: *A young man . . . "lest they give birth to monsters."* Thundy, *South Indian Folktales,* 116–17.

136: *Another Kadar folktale . . . inherited the kingdom.* Thundy, *South Indian Folktales,* 77–79.

137: *"The pleasure that . . . falling to the ground."* Varadarajaiyer, *Elephant in the Tamil Land,* 91.

VI

murals, monoliths, and miniatures

PAGE

146: *The earliest monastaries . . . and exquisite frescoes.* Vidya Dehejia, *Indian Art* (London: Phaidon, 1997), 112–13.

148: *On April 28, 1819 . . . sites in the world.* Ranjana Sengupta, *Ajanta and Ellora* (Hong Kong: The Guidebook Company, 1991), 16–17.

153: *In the eastern . . . on his shoulders. Valmiki Ramayana,* quoted in *The Life and Lore of the Elephant,* Robert Delort (London: Thames and Hudson, 1992), 152.

154: *Gajalakshmi represents . . . libations of elephants.* Heinrich Zimmer, *Myths and Symbols in Indian Art and Civilization* (Princeton: Princeton UP, 1946), 105–9 and K. S. Srivastava, *The Elephant in Early Indian Art* (Varanasi: Sangeeta Prakashan, 1989), 38–39.

155: *Elephants can be . . . presence of* Elephas maximus. Srivastava, *Elephant in Early Indian Art,* 11.

157: *An accompanying . . . pair of ducks.* B. N. Goswamy and Eberhard Fischer, *Wonders of a Golden Age* (Zurich: Museum Rietberg, 1987), 137.

157: *"droves of elephants . . . the elephant keepers."* Babur *Baburnama,* trans. Wheeler M. Thackston (New York: Oxford UP, 1996), 326.

157: *"They can easily . . . strings of camels." Baburnama,* 335.

157: *"When India was . . . is not said."* Abu-L-Fazl, *The Akbar Nama of Abu-L-Fazl,* trans. H. Beveridge (Delhi: Ess Ess Publications, 1977), 111.

158: *"without exaggeration . . . any army."* Abu-L-Fazl, *The Akbar Nama,* 116.

158: *"when the fumes . . . circulating in its brain."* Abu-L-Fazl, *The Akbar Nama,* 113.

158: *In a dramatic . . . into the river.* J. M. Rodgers, *Mughal Miniatures* (London: Thames and Hudson, 1994), 55.

158: *"One day, in the full bloom. . . . unsuitable for kings."* Jahangir, *Jahangirnama,* trans. Wheeler M. Thackston (New York: Oxford UP, 1999), 278–79.

158: *One particularly . . . the remaining victims . . .* Ashok Kumar Das, "The Elephant in Mughal Painting," in *Flora and Fauna in Mughal Art,* ed. Som Prakash Verma (Mumbai: Marg, 1999), 43.

159: *"When such marks . . . lower than fossils."* Abu-L-Fazl, *The Akbar Nama,* 490.

159: *Jahangir also had . . . compliments paid.* Jahangir, *Jahangirnama,* 93.

159: *"Of all animals . . . peculiarly mine."* Jahangir, *Jahangirnama,* 238.

160: *Aurangzeb is shown . . . filled with fireworks.* Inayar Khan, *The Shah Jahan Nama of Inayat Khan,* trans. W. E. Begley, (Delhi: Oxford UP, 1990), plate 20.

160: *"One day . . . ready to strike."* Inayar Khan, *The Shah Jahan Nama,* 95–96.

160: *A striking miniature . . . rescue of the elephants.* Rodgers, *Mughal Miniatures,* 88.

161: *Muhammad al-Damiri's . . . air onto ships.* al-Damiri, *Hayat al-Hayawan,* 856–57.

161: *The* Hastividyarnava *. . . gods at its root.* Sukumara Barkath,

Hastividyarnava, P. C. Choudhury, ed. (Gawahati: Publication Board Assam, 1976), viii.

162: *One of the objects . . . of a mirror.* Dehejia, *Indian Art,* 111.

162: *An elephant's tusk . . . fired by a hunter years ago.* George F. Kunz, *Ivory and the Elephant* (New York: Doubleday, 1916), 219–23.

163: *Since 1976 . . . a perverse value.* Menon, *Tusker,* 150, 225.

163–64: *In many Indian paintings . . . back of a tusker.* Robert J. Del Bonta, "Reinventing Nature: Mughal Composite Animal Paintings," in *Flora and Fauna in Mughal Art,* ed. Som Prakash Verma (Mumbai: Marg, 1999), 69–82 and Saller, *Elephants,* 125.

164–65: *Somewhat similar . . . literally become the text.* Kunz, *Ivory and the Elephant,* 167–68 and M. A. Ziaduddin, *Monograph on Moslem Calligraphy* (Lahore: Al-Biruni, 1979), 68.

165–66: *"However well . . . anatomical parts."* Stuart Carey Welch et al., *Gods, Kings, and Tigers* (New York: Prestel, 1997), 20.

166: *"An Angry Elephant Breaks Its Chains."* Welch et al., *Gods, Kings, and Tigers,* 108.

167: *The Chattar Mahal . . . the rhino's neck.* Welch et al., *Gods, Kings, and Tigers,* 16 (Fig. 1).

169: *"If they did not . . . also be fulfilled."* Zimmer, *Myth and Symbols in Indian Art,* 108.

VII
a r u n d h a t i 's   b a t h

PAGE

176: *Arundhati . . . banks of the Ganga.* John Dowson, *A Classical Dictionary of Hindu Mythology and Religion* (Delhi: Rupa, 1998), 24.

179: *"With the tape . . . range of human hearing."* Katy Payne, *Silent Thunder* (New York: Penguin, 1999), 28.

180: *glossary that appears . . . published in 1922.* A. J. W. Milroy, *Management of Elephants,* 99–100.

185: *Their research has shown . . . upstream from Hardwar.* A. J. T. Johnsingh et al., "Conservation of North-Western Elephant Range," unpublished (2001), 1–14 and A. J. T. Johnsingh and Justus

Joshua, "Conserving Rajaji and Corbett National Parks—The Elephant as a Flagship Species," *Oryx,* 28, no. 2 (1994), 135–40.

186–87: *"Thus ended the life . . . feel a sense of loss."* A. J. T. Johnsingh and A. Christy Williams, "Bull Elephants in the Rajaji-Corbett Range," *Sanctuary Asia* (June 2002): 43.

190: *"There is no bigger" . . . subject to discrepancies and human error.* Menon, *Tusker,* 39–46.

192: *Despite its picturesque beauty . . . forest department.* Anil Kumar Singh, "Elephant Mortality in Train Accidents" (Wildlife Trust of India, 2001), ii–v.

192: *Theories have . . . cannot hear.* Singh, "Elephant Mortality," 34.

192: *Rescheduling . . . dangers remain.* Singh, "Elephant Mortality," 27.

197: *These inscriptions . . . demands of this world.* Romila Thapar, *A History of India,* vol. 1 (Delhi: Penguin, 2000), 72.

VIII
gaja∂utra

PAGE

200: *Strategically located . . . in 493 B.C.* Thapar, *History of India,* vol. 1, 56.

200–1: *"The greatest city . . . admiration of the Greeks."* R. C. Mazumdar, *Ancient India* (Delhi: Motilal Banarasidass, 1971), 107.

201: *Pataliputra means . . . named it Pataliputra.* Hiuen Tsiang, *Buddhist Records of the Western World,* vol. 2, trans. Samuel Beal (London: Kegan Paul, Trench, Trubner & Co., 1968), 83–85.

201: *The kingdom of Magadha . . . "flame of the forest."* McCrindle, 205–6.

201: *In Hindi this tree . . . food of elephants.* K. C. Sahni, *The Book of Indian Trees* (Mumbai: BNHS, 1998), 87–88.

202: *"The attendants . . . sweet and pleasant."* J. W. McCrindle. *Ancient India as Described by Megasthenes and Arrian* (London: Trubner, 1877), 117–18.

202: *"Some of them . . . to the stupa."* Hiuen Tsiang, *Buddhist Records,* vol. 2, 28.

202: *Hiuen Tsiang ... compared to elephants.* Hiuen Tsiang, *Buddhist Records,* vol. 2, 138–39.

202–3: *In Buddhist iconography ... into the pillars at Sanchi.* V. S. Naravane, *The Elephant and the Lotus* (London: Asia Publishing House, 1965), 61.

203: *There is also ... shower her with water.* Naravane, *Elephant and the Lotus,* 9.

203: *The pleasure gardens ... named after Ashoka.* A. L. Basham, *The Wonder That Was India* (London, Sidgwick and Jackson, 1967), 204–205.

203: *Ashoka* (Saraca asoca) *... the god of love,* Sahni, 80–82.

203: *according to legend ... by a beautiful woman.* Basham, 204.

205: *Within the Mauryan ... sanctuaries of today.* Thomas Trautmann, "Elephants and the Mauryas," in *India: History and Thought,* ed. S. N. Mukherjee (Calcutta: Subarnarekha, 1982), 263–66.

205–6: *"The superintendent ... binders and others."* Kautilya, *Kautilya's Arthshastra,* trans B. K. Chaturvedi (Delhi: Diamond, 2001), 85.

206: *The precise dimensions ... "taking wakeful rest."* Kautilya, *Kautilya's Arthshastra,* 85.

206: *Kautilya lists ... "for the safety of the elephants."* Kautilya, *Kautilya's Arthshastra,* 86.

206–7: *"The superintendent ... dangerous to life."* Kautilya, *Kautilya's Arthshastra,* 31.

207: *Historian ... international arms trade.* Trautmann, "Elephants and the Mauryas," 254–59.

208: *"hunters tracking wild elephants."* Wendy Doniger O'Flaherty, *The Rig Veda* (London: Penguin, 1981), 265.

212: *"When one price is ... shall be arrived at."* Nilakantha, *Montangalila,* 73.

212: *In his book ... in Madhya Pradesh.* Mark Shand, *Travels on My Elephant* (London: Jonathan Cape, 1991), 182.

215: *Of the ninety-two ... symptoms of blindness.* N. V. K. Ashraf and Vidya Deshpande, "Report on the Elephant Health Camp, Sonpur Mela 2001" (Delhi: WTI, 2002), 11.

217: *An elephant king ... into a crocodile.* Shakti M. Gupta, *Myths and*

Legends from Indian Mythology (Delhi: B. R. Publishing, 2002), 169–71.

220–21: *"Indian women . . . worth an elephant."* McCrindle, 222, also cited in Trautmann, "Elephants and the Mauryas," 254.

221: *"Who art thou . . . trunk of an elephant."* Dimock et al., *Literatures of India,* 65.

221–22: *Without question . . . each amorous couple.* Vatsyayana, *The Kamasutra,* trans. and eds. Wendy Doniger and Sudhir Kakar (New York: Oxford, 2002), xi–xxv.

222: *Varying dimensions . . . genital hierarchy.* Vatsyayana, 28–39.

223: *"A man who . . . wild animals, and birds."* Vatsyayana, 56.

227–28: *"see the problem . . . ecological harmony."* Madhav Gadgil and Ramachandra Guha, *Ecology and Equity* (Delhi: Penguin, 1995), 108.

IX

p o w e r   a n d   p o m p

PAGES

235–36: *In recent years . . . contraception.* I am indebted to Jeffrey Campbell for this story, details of which were confirmed in Oscar Harkavy, *Curbing Population Growth: An Insider's Perspective* (New York: Plenum Press, 1995), 138.

237: *"One may ask . . . of the Shah."* Simon Digby, *War-Horse and Elephant in the Delhi Sultanate* (Oxford: Orient Monographs, 1971), 56.

237–38: *During this conflict . . . power in Hastinapur.* C. Rajagopalachari, *Mahabharata* (Mumbai: Bharatiya Vidya Bhavan, 1999), 380–84.

239: *The number of elephants . . . against Alexander.* Trautmann, "Elephants and the Mauryas," 266.

239: *The Mughal . . . used in battle.* Digby, *War-Horse and Elephant,* 57.

239: *A hundred years earlier . . . 120 war elephants.* Digby, *War-Horse and Elephant,* 59.

239: *"The rank of elephants . . . upon the mountain"* Digby, *War-Horse and Elephant,* 53.

240: *"From arrows . . . his life fled."* Digby, *War-Horse and Elephant,* 53.

240: *"The great reliance . . . its own side."* Basham, 130.

240–41: *"The elephant is a . . . the horse, evoked."* Digby, *War-Horse and Elephant,* 52.

241: *In one dramatic . . . "of all the elephants."* Digby, *War-Horse and Elephant,* 77–79.

241–42: *By the time of . . . these soon disappeared.* Digby, *War-Horse and Elephant,* 65–66.

243: *"The elephants . . . away altogether."* Digby, *War-Horse and Elephant,* 51.

244: *In 1903 . . . the Victorian era.* Percival Spear, *A History of India,* vol. 2, 1965 (London, Pelican, 1973), 174–75.

245: *"The torn boughs . . . of the queen."* John Lockwood Kipling, *Beast and Man in India,* 207.

245: *"The use of elephants . . . of the Raj."* Lahiri-Choudhury, *Great Indian Elephant Book,* xxi.

245: *"For centuries . . . possibly have achieved."* Mortimer and Dorothy Menpes, *The Durbar* (London: Charles Black, 1903), 203–04.

246: *"glittering gold umbrella."* Menpes, *Durbar,* 45.

246: *The only person . . . entire procession.* Lahiri-Choudhury, *Great Indian Elephant Book,* xxii.

246–47: *However, elephants did serve . . . Survey of India.* John Keay, *The Great Arc* (London: HarperCollins, 2000), 5.

247: *"Without them . . . more difficult."* Sir William Slim, "Foreword," in *Elephant Bill,* J. H. Williams (London: Rupert Hart-Davis, 1955), n.p.

247: *One of the many . . . side by side.* Williams, *Elephant Bill,* 148–49.

248–49: *"And it was . . . that he destroys."* George Orwell, "Shooting an Elephant," *Shooting an Elephant and Other Essays* (New York: Harcourt, 1945), 8.

249: *"sagged flabbily . . . damage him further."* Orwell, "Shooting an Elephant," 10.

249: *"I often wondered . . . avoid looking a fool."* Orwell, "Shooting an Elephant," 12.

249: *"the greatest hunter of modern times."* Harriet Ritvo, *The Animal Estate* (Cambridge, Harvard, 1987), 250.

251: *"Ah, comrade . . . sahibs are great!"* A. Mervyn Smith, "Peer Bux: The Terror of Hunsar," in *Great Indian Hunting Stories,* ed. Stephen Alter (Delhi: Penguin, 1988), 93.

252: *Project Elephant . . . captive animals.* S. S. Bist, "Introduction," in *Management of Elephants in Captivity,* A. J. W. Milroy (Dehradun: Natraj, 2002), i.

252: *Figures like these . . . clearly outnumbered.* Menon, *Tusker,* 39–40.

254: *"Domesticated elephants . . . conservation imperative."* Bist, "Introduction," iii.

X

n o r t h e a s t

260: *"O my mahout . . . on a tusker."* Mark Shand, *Elephant Queen* (London: Jonathan Cape, 1995), 182.

260: *"O girl from Gauripur . . . a wild elephant than this girl."* Bupen Hazarika, "O Girl from Gauripur," recited and translated by Hemanta Das.

261: *"suitable norms . . . of the law."* Parbati Barua and S. S. Bist, "Cruelty to Elephants—A Legal and Practical View," *Zoo's Print* (June 1996): 51.

261: *She has defended . . . forest department official.* Parbati Barua, "Facts on Capture and Death of a Wild Elephant in Chattisgarh," http://www.indianjungles.com/290303.htm.

262: *"With their capacity . . . medicinal herbs."* Parbati Barua, "Image and Profession of Mahouts in North Eastern India." *Zoo's Print* (June 1996): 34.

263–64: *Many years . . . Ganjang was dead.* Dewansing Ronmitu Sangma, *Jadoreng* (Shillong: Salsang C. Marak, 1993), 275–76.

264–65: *Another shaman . . . was confirmed.* Sangma, *Jadoreng,* 267–69.

269–70: *Vivek Menon . . . ("Rice Thief, Bin Laden!")* Menon, *Tusker,* 50–51.

270: *Out of a total area . . . in the crossfire.* Wasbir Hussain, "Demolition Day," *India Today,* 8 July 2002, 50–52.

275: *This book shares . . . word for word.* Barkath, *Hastividyarnava,* iv–v, 8.

276: *These riverine cetaceans . . . hundred teeth.* Prater, *Book of Indian Animals,* 313–14.

278: *These are made . . . used as incense.* Sahni, *Book of Indian Trees,* 142–44.

280: *"make the forests . . . because of such elephants."* Barkath, *Hastividyarnava,* 32.

284: *"One of the . . .* terra incognita.*"* E. P. Gee, *Wild Life of India,* 180–81.

284: *"No one can . . . cannot go there."* Gee, *Wild Life of India,* 181.

285: *"Wildlife? . . . far too much."* Jawaharlal Nehru, "Foreword," in *The Wild Life of India,* E. P. Gee, 5–6.

287: *"a closely-matted . . . from the skin."* Prater, *Book of Indian Animals,* 229.

287: *Poaching and the . . . depend for food.* Wasbir Hussain. "Exotic No More," *India Today,* 3 June 2002, 46–47.

288: *"studded with masses of rounded tubercles."* Prater, *Book of Indian Animals,* 230.

bibliography

Abu-L-Fazl. *Abu-L-Fazl's Akbar Nama.* Trans. H. Beveridge. Vol. 1–3. Delhi: Ess Ess, 1971.

al-Damiri, Muhammad. *Hayat al-Hayawan.* Trans. A. S. G. Jayakar. Bombay: D. B. Taraporevala, 1906.

Ali, Salim. *The Fall of a Sparrow.* Delhi: Oxford UP, 1985.

"Allegations Regarding Cruel Treatment of an Elephant." Press Release. Embassy of India. Washington, D.C. http://www.indianembassy. org/pic/PR_1999/March99/prmarch1799.html.

Ashraf, N. V. K., and Vidya Deshpande. *Report on the Elephant Health Camp, Sonpur Mela 2001.* Delhi: WTI, 2002.

Babur. *The Baburnama.* Trans. Wheeler M. Thackston. New York: Oxford UP, 1996.

Barkath, Sukumara. *Hastividyarnava.* Choudhury, Pratap C., ed. Guwahati: Publication Board Assam, 1976.

Barua, Parbati, and S. S. Bist. "Cruelty to Elephants—A Legal and Practical View." *Zoo's Print* June 1996: 47–51.

Barua, Parbati. "Elephant Capturing in North-Eastern India." *Zoo's Print* June 1996: 33–34.

———. "Facts on Capture and Death of a Wild Elephant in Chattisgarh." http://www.indianjungles.com/290303.htm.

———. "Image and Profession of Mahouts in North Eastern India." *Zoo's Print* June 1996: 34.

————. "Sick and Injured Elephants: Care and Cure." *Zoo's Print* June 1996: 21–23.

Basham, A. L. *The Wonder That Was India.* London: Sidgwick and Jackson, 1967.

Baskaran, N. et al. "Home Range of Elephants in the Nilgiri Biosphere Reserve, South India." *A Week With Elephants.* Eds. J. C. Daniel and H. S. Datye. Mumbai: BNHS, 1995. 296–313.

————. "Crop Raiding by Asian Elephants in the Nilgiri Biosphere Reserve, South India." *A Week With Elephants.* Eds. J. C. Daniel and H. S. Datye. Mumbai: BNHS, 1995. 350–67.

Brown, Robert L., ed. *Ganesh: Studies of an Asian God.* Delhi: Sri Satguru, 1997.

Corbett, Jim. "On Man-eating." *Jim Corbett's India.* Ed. R. E. Hawkins. Delhi: Oxford UP, 1978.

————. *The Temple Tiger.* London: Oxford UP, 1955.

Daniel, J. C. "Conservation and Future of the Indian Elephant." Inaugural Address at the Conference on "The Call of the Elephant." Indian Museum, Kolkata, 18–19 August 2001.

———— and Hemant Datye. *A Week with Elephants.* Mumbai: BNHS, 1995.

————. *The Asian Elephant: A Natural History.* Dehradun: Natraj, 1998.

Darwin, Charles. *The Expression of the Emotions in Man and Animal.* Vol. 23 in *The Works of Charles Darwin.* Eds. Paul H. Barrett and R. B. Freeman. London: William Pickering, 1989.

————. *On the Origin of Species.* New York: Mentor, 1958.

Das, Asok Kumar. "The Elephant in Mughal Painting." *Flora and Fauna in Mughal Art.* Ed. Som Prakash Verma. Mumbai: Marg, 1999. 36–54.

Dehejia, Vidya. *Indian Art.* London: Phaidon, 1997.

Del Bonta, Robert J. "Reinventing Nature: Mughal Composite Painting." *Flora and Fauna in Mughal Art.* Ed. Som Prakash Verma. Mumbai: Marg, 1999. 69–82.

Delort, Robert. *The Life and Lore of the Elephant.* Trans. I. Mark Paris. London: Thames and Hudson, 1992.

Dhavalikhar, M. K. "Ganesa: Myth and Reality." *Ganesh: Studies of an Asian God.* Ed. Robert L. Brown. Delhi: Sri Satguru, 1997. 49–61.

Digby, Simon. *War-Horse and Elephant in the Delhi Sultanate.* Oxford: Orient Monographs, 1971.

Dimock, Edward C. et al. *The Literatures of India: An Introduction.* Chicago: University of Chicago Press, 1978.

Dowson, John. *A Classical Dictionary of Hindu Mythology and Religion.* Delhi: Rupa, 1998.

Elephant Boy. Dir. Robert Flaherty. London: London Films, 1937.

Flaherty, Frances. *Elephant Dance.* New York: Scribners, 1937.

"Forgotten Hollywood Star from India." *The Hindu* 18 February 1990, Bangalore ed.: 20.

Fox, Michael. "New Problems, May 2002." http://www.gcci.org/ipan/loki/newproblems0502.html.

————. "The Worst Case of Animal Abuse I Have Ever Documented." IPAN Press Release. 1 March 1999. http://www.gcci.org/ipan/loki/press030199.

Gadgil, Madhav, and Ramachandra Guha. *Ecology and Equity: The Use and Abuse of Nature in Contemporary India.* Delhi: Penguin, 1995.

Gaeth, A. P. "The Developing Renal, Reproductive, and Respiratory Systems of the African Elephant Suggest an Aquatic Ancestry." *Journal of Developmental Biology* 96, no. 10 (1999). http://www.pnas.org.

Gandhi, Maneka. "Retyped Letter from Maneka Ghandhi (sic) to English Woman Who Visited IPAN and Has Been Enquiring About Condition of Loki, the Elephant." 7 September 1999. http://www.gcci.org/ipan/loki/maneka_ghandi090799.html.

Gee, E. P. *The Wild Life of India.* London: Fontana, 1964.

George, Jacob. "Trunk Call." *India Today* 24 February 2003: 42.

Goswamy, B. N., and Eberhard Fischer. *Wonders of a Golden Age.* Zurich: Museum Rietberg, 1987.

Gupta, S. K. *Elephant in Indian Art and Mythology.* Delhi: Abhinav, 1983.

Gupta, Shakti M. *Myths and Legends from Indian Mythology.* Delhi: B. R. Publishing, 2002.

Guruvayur Kesavan. Dir. Bharatan. Kerala: Rohit Video, 1977.

Harkavy, Oscar. *Curbing Population Growth.* New York: Plenum Press, 1995.

Hathi Mera Sathi. Dir. M. A. Thirumugam. Madras: Devar Films, 1971.

Hussain, Wasbir. "Demolition Day." *India Today* 8 July 2002: 50–52.

———. "Exotic No More." *India Today* 3 June 2002: 46–47.

Jaganathan, Shakunthala, and Nanditha Krishna. *Ganesha.* Mumbai: Vakils, 2000.

Jahangir. *The Jahangirnama.* Trans. Wheeler M. Thackston. New York: Oxford UP, 1999.

Johnsingh, A. J. T. "Rajaji." *Sanctuary* 11 (1991): 14–25.

Johnsingh, A. J. T., and A. Christy Wilson. "Bull Elephants in the Rajaji-Corbett Range." *Sanctuary Asia* June 2002: 38–45.

———. "Elephant Corridors in India: Lessons for Other Elephant Range Countries." *Oryx* 33.3 (1999): 210–14.

Johnsingh, A. J. T., and Justus Joshua. "Conserving Rajaji and Corbett National Parks—The Elephant as Flagship Species." *Oryx* 28.2 (1994): 135–40.

Johnsingh, A. J. T., S. N. Prasad, and S. P. Goyal. "Conservation Status of the Chila-Motichur Corridor for Elephant Movement in Rajaji-Corbett National Parks Area, India." *Botanical Conservation* 51 (1990): 125–38.

Joshi, Poornima. "Slow, Painful Death at Corbett." *The Hindustan Times* 21 January 2001: 1.

Kalidasa. *The Megha-Duta of Kalidasa.* Trans. Sushil Kumar De. Delhi: Sahitya Akademi, 1957.

———. *The Meghaduta of Kalidasa.* Trans. M. R. Kale. Delhi: Motilal Banarasidass, 1991.

———. *The Meghaduta of Kalidasa.* Trans. G. H. Rooke. London: Oxford UP, 1935.

———. *The Megha Duta or Cloud Messenger.* Trans. Horace H. Wilson. London: Black, Parry, and Co., 1814.

Kautilya. *Kautilya's Arthashastra.* Trans. B. K. Chaturvedi. Delhi: Diamond, 2001.

Kaye, Myriam. *An Illustrated Guide to Bombay and Goa.* Hong Kong: The Guidebook Co., 1990.

Keay, John. *The Great Arc.* London: HarperCollins, 2000.

Khan, Inayat. *Shah Jahan Nama.* Trans. A. R. Fuller. Delhi: Oxford UP, 1990.

Kipling, John Lockwood. *Beast and Man in India.* London: Macmillan, 1921.

Kipling, Rudyard. "Toomai of the Elephants." *The Jungle Book.* New York: The Century Company, 1899.

Kolhatkar, Neeta. "Kal Aaj Aur Kal." *Times of India* 29 August 2001, special supplement: 1.

———. "Tilak Revisits Girgaum!" *Times of India,* 30 August 2001, special supplement: 1.

Krishnan, M. "Musth." *The Asian Elephant: A Natural History.* Ed. J. C. Daniel. Dehradun: Natraj, 1998: 126–28.

Kunz, George F. *Ivory and the Elephant.* New York, Doubleday, 1916.

Lahiri-Choudhury, D. K. "The Call of the Elephant." Keynote Address at the Conference on "The Call of the Elephant." Indian Museum, Kolkata, 18–19 August 2001.

———. *The Great Indian Elephant Book.* Delhi: Oxford University Press, 1999.

"Loki." Encyclopedia Mythica, http://www. pantheon.org/articles/l/loki.html.

"Loki." http://www.islandia.is/~oldnorse/gods/loki.htm.

Lopez, Barry. *Arctic Dreams.* New York: Bantam, 1986.

Majumdar, R. C. *Ancient India.* 6th ed. Delhi: Motilal Banarasidass, 1971.

Marak, Julius L. R. *Balpakram: The Land of Spirits.* Delhi: Akansha, 2000.

Mariappa, D. *Anatomy and Histology of the Indian Elephant.* Oak Park: Indira, 1986.

McCrindle, J. W. *Ancient India as Described by Megasthenes and Arrian.* London: Trubner & Co., 1877.

Menon, Vivek. *Tusker: The Story of the Asian Elephant.* Delhi: Penguin, 2002.

Menpes, Mortimer and Dorothy. *The Durbar.* London: Charles Black, 1903.

Miall, L. C., and F. Greenwood. *Anatomy of the Indian Elephant.* London: Macmillan, 1878.

Monier-Williams, Monier. *A Sanskrit-English Dictionary.* Delhi: Asian Educational Services, 1999.

Nagar, Shanti Lal. *Jatakas in Indian Art.* Delhi: Parimal, 1993.

Narain, A. K. "Ganesa: A Protohistory of the Idea and the Icon." *Ganesa: Studies of an Asian God.* Ed. R. L. Brown. Delhi: Sri Satguru, 1997. 19–48.

Naravane, V. S. *The Elephant and the Lotus.* London: Asia Publishing, 1965.

Narayan, R. K. *The Ramayana.* New York: Penguin, 1972.

Nilakantha. *Elephant Lore of the Hindus (Matanga-lila by Nilakantha).* Trans. Franklin Edgerton. New Haven: Yale UP, 1931.

Orwell, George. "Shooting an Elephant." *Shooting an Elephant and Other Essays.* New York: Harcourt, 1945. 3–12.

Panchatantra, The. Trans. Arthur W. Ryder. Chicago: University of Chicago Press, 1926.

Payne, Katy. *Silent Thunder.* New York: Penguin, 1999.

Physiologus. Trans. Michael J. Curley. Austin: University of Texas Press, 1979.

Pliny the Elder. *Natural History (Naturalis Historia).* Trans. H. Rakham. Cambridge: Harvard UP, 1940.

"Poachers Kill Ranger, Injure Three Guards in Corbett." *Times of India* 30 August 2001: 7.

Prater, S. H. *The Book of Indian Animals.* Mumbai: BNHS, 1948.

Raghuram, Sunaad. *Veerappan: The Untold Story.* Delhi: Penguin, 2001.

Rajagopalachari, C. *Mahabharata.* Mumbai: Bharatiya Vidya Bhavan, 1999.

———. *Ramayana.* Mumbai: Bharatiya Vidya Bhavan, 1996.

Rao, C. Hayavadana. *Mysore Gazetteer.* Bangalore: Government Press, 1927.

Rao, M. Shama. *Modern Mysore.* Bangalore: Higginbothams, 1936.

Rathod, Kishore. "51 Ways Modaks Can Delight." *Sunday Mid-Day,* 26 Aug. 2001: 5.

Rig Veda, The. Wendy Doniger O'Flaherty. New York: Penguin, 1981.

Ritvo, Harriet. *The Animal Estate.* Cambidge: Harvard University Press, 1987.

Rodgers, J. M. *Mughal Miniatures.* London: Thames and Hudson, 1994.

Sahni, K. C. *The Book of Indian Trees.* Mumbai: BNHS, 1998.

Saller, Martin, and Karl Groning. *Elephants: A Cultural and Natural History.* Cologne: Konemann, 1999.

Sanderson, G. P. *Thirteen Years Among the Wild Beasts of India.* London: W. H. Allen, 1879.

Sangma, D. R. *Jadoreng (The Psycho-Physical Culture of the Garos).* Shillong: S. C. Marak, 1993.

Scigliano, Eric. *Love, War, and Circuses: The Age-Old Relationship Between Elephants and Humans.* Boston: Houghton Mifflin, 2002.

Sengupta, Ranjana. *Ajanta and Ellora.* Hong Kong: The Guidebook Co., 1991.

Shand, Mark. *Elephant Queen.* London: Jonathan Cape, 1995.

————. *Travels on My Elephant.* London: Jonathan Cape, 1991.

Singh, Anil Kumar. *Elephant Mortality in Train Accidents: A Scientific Approach to Understanding and Mitigating This Problem in Rajaji National Park.* Delhi: Wildlife Trust of India, 2001.

Singh, Brijraj. *Festivals and Ceremonies Observed by the Royal Family of Kotah.* Zurich: Museum Rietberg, 1998.

Smith, A. Mervyn. "Peer Bux: The Terror of Hunsar." *Great Indian Hunting Stories.* Ed. Stephen Alter. Delhi: Penguin, 1988. 79–93.

Spear, Percival. *A History of India.* Vol 2. London: Pelican, 1973.

Sridhar, R. "Make Way for the Lalbaughcha Raja." *Times of India* 30 Aug. 2001, supplement: 3.

Srivastava, Kamal Shankar. *The Elephant in Early Indian Art.* Varanasi: Sangeeta Prakashan, 1989.

Stracey, P. D. *Elephant Gold.* London: Weidenfeld and Nicholson, 1963.

Sukumar, R. *The Asian Elephant: Ecology and Management.* Cambridge: Cambridge University Press, 1989.

———. *Elephant Days and Nights: Ten Years With the Indian Elephant.* Delhi: Oxford, 1994.

———. "Letter to IPAN." 22 April 1999. http://www.gcci.org/ipan/loki/raman_sukumar042299.html.

Swarup, Shanti. *Flora and Fauna in Mughal Art.* Mumbai: Taraporevala, 1983.

Thapar, Romila. *A History of India.* Vol. 1. Delhi: Penguin, 2000.

Thundy, Zacharias P. *South Indian Folktales of Kadar.* Meerut: Archana, 1983.

Trautmann, Thomas R. "Elephants and the Mauryas." *India: History and Thought.* Ed. S. N. Mukherjee. Kolkata: Subarnarekha, 1982.

Tsiang, Hiuen. *Buddhist Records of the Western World.* Trans. Samuel Beal. London: Kegan Paul, Trench, Trubner & Co., 1968.

Varadarajaiyer, E. S. *The Elephant in the Tamil Land.* Annamalainagar: Annamalai University (Tamil Series No. 8), 1945.

Vatsyayana. *Kamasutra.* Trans. and Eds. Wendy Doniger and Sudhir Kakar. New York: Oxford UP, 2002.

Verma, Som Prakash. *Flora and Fauna in Mughal Art.* Mumbai: Marg, 1999.

Welch, Stuart Cary et al. *Gods, Kings, and Tigers: The Art of Kotah.* New York: Prestel, 1997.

Williams, J. H. *Elephant Bill.* London: Rupert Hart-Davis, 1955.

Winchester, Simon. *The Professor and the Madman.* New York: Perennial, 1999.

Ziaduddin, M. *A Monograph on Moslem Calligraphy.* Lahore: Al-Biruni, 1979.

Zimmer, Heinrich. *Myths and Symbols in Indian Art and Civilization.* Ed. Joseph Campbell. Princeton: Princeton University Press, 1946.